Gabriel Mendes de Almeida Carvalho
Alexandre Bracarense
Ezequiel Caires Pereira Pessoa

Study of techniques for improving wet underwater welds

Gabriel Mendes de Almeida Carvalho
Alexandre Bracarense
Ezequiel Caires Pereira Pessoa

Study of techniques for improving wet underwater welds

Imprint

Any brand names and product names mentioned in this book are subject to trademark, brand or patent protection and are trademarks or registered trademarks of their respective holders. The use of brand names, product names, common names, trade names, product descriptions etc. even without a particular marking in this work is in no way to be construed to mean that such names may be regarded as unrestricted in respect of trademark and brand protection legislation and could thus be used by anyone.

Cover image: www.ingimage.com

This book is a translation from the original published under ISBN 978-613-9-64823-8.

Publisher:
Sciencia Scripts
is a trademark of
Dodo Books Indian Ocean Ltd. and OmniScriptum S.R.L publishing group

120 High Road, East Finchley, London, N2 9ED, United Kingdom
Str. Armeneasca 28/1, office 1, Chisinau MD-2012, Republic of Moldova, Europe
Printed at: see last page
ISBN: 978-620-7-78617-6

Dedicated to all those who seek every day to become better at what they set out to do.

ACKNOWLEDGEMENTS

I always start my thanks to the Being who I believe is above our energies here on Earth, no matter what the name, He gives us gifts and abilities to achieve goals and have good times in this life.

Thank you God for the physical, mental and emotional ability to achieve this dream.

To my parents Elvio and Socorro, for their investment in me as a son and a citizen, always based on their immeasurable love for me and mine for them;

To my brothers Guilherme and Rafael, for their companionship and for being the greatest moulders of my being.

To my family for their prayers and constant support, thank you for always being my source of peace and calm.

To my advisor, Prof Alexandre Queiroz Bracarense, for overcoming the advisor-student barrier and for being by my side in yet another stage of my constant quest to one day surpass him.

To my co-supervisor, Prof Ezequiel Caires Pereira Pessoa, for introducing me to underwater welding and for guiding me through the research.

To Prof Paulo José Modenesi for his teaching in the classroom and his availability outside it.

To my friends and daily work mates at the Laboratory (LRSS) for their encouragement and patience. Special thanks to Carol, Nilo, Gedael, Frank, Mateusão, Mateus Cip, Fagner, Prof Ariel, Carolzinha and Bruno, Camila, Café, Prof Cláudio, Danielle, Diego, Diogo, Fernando, Filipe, Henrique, José Roberto, Leandro bombeiro, Luana, Luciano, Luiz, Marcelo, Marco, Moara, Renata, Sheron, Tarcísio, the journey would not have been the same without you.

To Andrejito and Miguelito for their weekly "peleas", for their companionship in and out of the house and for their Spanish lessons.

To my fellow article writers, Arthur, Bruno and Rafael, may the work continue.

Thanks to UFMG for the structure and funding, especially Marina and Rayanne, for sparing no effort to ensure the smooth running of the postgraduate programme.

To CAPES and PETROBRAS for the funding.

My sincere thanks.

"Never let anyone tell you
That it's not worth believing in a dream you have,
That your plans will never work out
Or that you'll never amount to anything.
(Renato Russo)

SUMMARY

SUMMARY

Structures used in the offshore sector, such as ships and platforms, are designed to withstand a wide variety of stresses while remaining intact and, above all, safe. In order to meet these needs for longer, the maintenance carried out on them must maintain the required mechanical properties. For economic reasons, the maintenance of these structures is almost always carried out offshore. Wet welding with coated electrodes (SMER) has been widely used due to its low cost compared to other processes, its versatility in work and the fact that it already achieves good quality results. Despite the improvements achieved in recent years, especially in the quality of the metal deposited, there are still problems in maintaining the mechanical properties at adequate values in the "welding region". One of the main problems is the loss of ductility in heated regions adjacent to the weld or the thermally affected zone (HAZ). Due to the rapid cooling inherent in the wet process, a microstructure of high hardness and low toughness is generated. A possible alternative for improving toughness in these regions is to apply a post-weld heat treatment (PWT), as has already been studied in non-wet processes. However, the application of a TTPS is sometimes unfeasible and difficult to implement in the underwater environment. An alternative way of improving the properties of multi-pass welds could be to use the heat from subsequent passes to temper hardened areas and thus achieve adequate levels of hardness.

This work studied the effect of two techniques and combinations of them during underwater welding without TTPS, to control the hardness of the HAZ of multipass wet welds. One of the techniques used was grinding to roughen the bead with each deposited weld pass, always leaving a reinforcement Imm. The other technique used was to reverse the welding direction at each deposition layer. For the study, a chamfer was developed with an opening close to the diameter of the electrode used and with parallel walls, so that the deposition would be better behaved, also reflecting a better positioning of the HAZ to be evaluated. The material chosen was SAE 1045 steel, as it is water-hardenable and has regions of high hardness when cooled quickly. The electrode used was E 6013, suitably varnished for underwater welding. All the welds were carried out in a hyperbaric camera adapted to simulate deep welds. The welds were made at a pressure equivalent to 10 metres deep.The results showed that when no technique is applied, internal points of the HAZ present regions of high hardness in which the heat from subsequent passes is not sufficient to promote adequate tempering of the material. However, when the grinding technique is used, removing part of the reinforcement, the heat is better utilised and the tempering achieved promotes softening of the entire length of the HAZ, leaving acceptable hardness values below the limits set by the standard. On the other hand, no improvement was noted when using the inversion technique in the welding direction.

Keywords: Underwater welding; UWW; Offshore welding.

CHAPTER 1

INTRODUCTION

The joining of metals by welding can present problems of metallurgical origin due to the composition of the material and the thermal cycle imposed at the time of welding. Phenomena inherent to the process include grain growth, the appearance of brittle structures, cracks, among others (CHOPRA et al. 2006), which can lead to failures.

Many difficulties of metallurgical origin are encountered in the welding of hardenable and thicker steels, and the structural fabrication codes (ASME, 1995) recommend carrying out post-weld heat treatment for these steels, usually at temperatures above 600°C. The main reasons for specifying TTPS are to improve the metallurgical structure and properties of welded joints, reduce the risk of brittle fracture failure, stress corrosion cracking, relieve residual stresses and improve fatigue performance (AGUIAR, 2001).

In many cases it is not feasible to apply a TTPS, especially when you need to repair an already assembled structure, a large component or carry out the procedure in an area that is difficult to access. One of the most complex cases for applying a TTPS is when welding underwater components, which is the scope of this work.

Underwater wet welds are classified according to the AWS D3.6 standard (AWS, 2010) into class A and B welds, and several class B procedures have been approved in Brazil for wet welds in recent decades (PESSOA, et al., 2013). Attempts to approve class A welding procedures have failed in hardness measurements in the HAZ, whose limit is set by the standard at 325HV for structural steels (AWS, 2010).

The high hardnesses found in the HAZ of wet welds are the result of the use of steels with high hardenability, normally found in oil platform structures, in combination with the high cooling rates (BRACARENSE, et al., 2008) found in wet welding.

The literature reports the application of some procedures aimed at reducing hardness in the HAZ of wet welds, such as the application of post-weld heat treatment using a gas or oxy-hydro flame (IBARRA & SZELAGOWSKI, 1992) and other procedures that do not fall under TTPS, such as: increasing the heat input (PESSOA, et al., 2013), applying a laser (FUKUDA, et al., 2009) and controlling the cooling rate (ZHANG, DA, FUKUDA, et al., 2013), 2013), application of the tempering pass (OLSEN, 1982), laser application (FUKUDA, et al., 2009) and control of the cooling rate (ZHANG, DAI, FENG, & HU, 2015); (TSAI & MASUBUCHI, 1979); (HASUI & SUGA, 1980).

All of these techniques are very difficult to apply in the field in the maritime conditions common on the Brazilian coast, whether in terms of control, positioning or parameterisation. These difficulties lead to a lack of repeatability and guarantee of effective control of the hardness of wet weld HAZs.

Therefore, there are strong technical and economic motivations that explain the major international effort undertaken since the late 1980s to develop repair procedures without TTPS that guarantee acceptable mechanical properties and good structural performance.

There are many techniques that use the energy from the multipass welding process to improve the mechanical properties of welded joints, including the half-layer technique, the alternating layer technique, the consistent layer technique and the controlled deposition technique (KÚCHLER, 2009).

Despite advances in techniques, there are still difficulties in choosing the right criteria for welding and a lack of data on the in-service performance of repaired components (SPERKO, 2005). The difficulties are even greater when the welding process is underwater and in the presence of water.

Based on the information presented, the aim of this work is to report and discuss the influence of applying simple techniques without subsequent heat treatment, which aim to improve the mechanical properties of welded joints in the presence of water, especially in controlling excessive hardness. This work will study the techniques of **"grinding the reinforcement at each pass"** and **"reversing the welding direction",** both for multi-pass welds.

The grinding technique is expected to make better use of the thermal cycles imposed in multipass welding, with the aim of softening the HAZ. With the direction reversal technique, there is a change in the direction of grain growth in the molten zone, which can improve the properties of the weld bead.

The welds were carried out in a hyperbaric tank, simulating the conditions found in the field, such as the presence of water and the pressure exerted by the environment. SAE 1045 medium carbon steel and the coated electrode process (SMAW) were used to study the influence of the proposed procedures on the properties of the HAZ.

The specific objectives of this work were:

Encourage research to improve the properties of components welded in underwater welding processes and propose new studies, especially on heat flow in welding in the presence of water.

CHAPTER 2
LITERATURE REVIEW

2.1 Underwater Welding

Until the emergence of offshore oil extraction, underwater welding was only used for emergency repairs to ship hulls and submerged harbour structures. The drastic increase in the price of a barrel of oil in 1973 meant that oil extraction, which had been concentrated almost entirely on land, expanded rapidly to the sea coast. Consequently, the need for welding repairs to these structures has stimulated a great deal of research in this area (FILHO *et al.*, 2004).

There are many end-of-life platforms in the world today, but because they are economically or strategically important for companies, they cannot be removed. Many of them are therefore being upgraded and reinforced.

To repair damage or reinforce platforms, reliable repair techniques must be used, including mechanical clamps, dry welding in hyperbaric chambers and wet welding.

Underwater welding can be subdivided into wet welding, local cavity welding and dry welding, as shown in Figure 2.1 (CHRISTENSEN, 1983). There are five types of technique in use: lat welding, habitat welding, dry chamber welding, dry spot welding and wet welding (AWS, 2011).

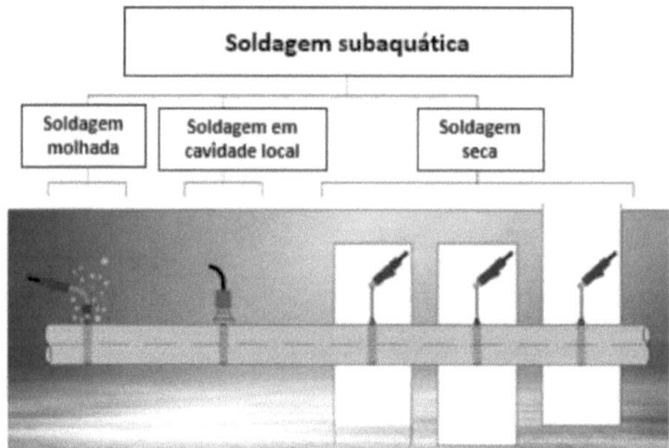

FIGURE 2.1 - Classification of underwater welding techniques.
SOURCE - LABANOWSKI et al., 2008, p.12.

The advantages of dry underwater welding lie in the fact that the dry environment promotes improvements in the stability of the electric arc, reduces problems with hydrogen in the weld metal, results in a lower rate of hardening of the base metal, and allows the restoration of the weld properties of ductility and toughness, which are the main problems of welds produced in wet underwater welding. As a disadvantage, dry welding requires a high investment value in the chamber and apparatus for removing water from the environment, which is not necessary in wet welding (AWS, 2011).

22   Wet Underwater Welding Processes

Wet welding is a method in which the joint to be welded is in direct contact with water. According to SILVA (2013), wet welding is a very complex process because it is carried out directly in contact with the aqueous medium. Because of this, problems arise that affect the mechanical properties of the weld metal, such as the high cooling rate and the dissociation of water (into H+ and OH-) due to the heat generated by the electric arc. Wet welding is the simplest technique, in which the joint and the diver are in contact with the ambient water without any physical barrier between the water and the electric arc (AWS, 2011; SILVA, 1983).

Wet welding developed from the 1940s, when moisture-resistant consumables were developed (KEATS, 2005). This technique is mainly used for temporary repairs to ships (GARASIC et al., 2010), oil platforms and pipelines and in cases of emergencies and accidents (ANAND and KHAJURIA, 2013).

The material most welded in wet conditions is steel (CARY and HELZER, 2005). Among the welding processes, the most commonly used is SMAW, followed by FCAW (AWS, 2011), but friction welding can also be used (GRANJON, 1983). The advantages that justify the preference for the SMAW process are the simplicity of the equipment (power source, cables, electrode holder and waterproof electrodes), low cost, mobility, ease of use and speed of repair, which allow it to be applied in places with unstable climates (WERNICKE, 1998).

The execution of underwater wet welding repairs on oil platform structures operating at high depths is still quite restricted, since these materials have high mechanical strength (oe>350MPA) and consequently high carbon equivalent, which contributes significantly to the occurrence of hydrogen cracking (FILHO *et al.*, 2004).Research has been carried out with the aim of solving the inherent problems of wet welding, excessive porosity, mechanical properties inferior to air welds and large amounts of hydrogen in the weld metal. As a result of these developments, wet welding has been successfully applied to the repair of platforms in the Gulf of Mexico (abroad, numerous interventions for structural repairs using underwater wet welding on platforms have been reported, notable among which are those carried out on fixed platforms in the Gulf of Mexico (PESSOA, 2007).

23   Wet underwater welding with coated electrodes

Wet welding with a coated electrode has its main application in the repair of structures, used mainly in the shipbuilding, gas and oil exploration industries and, more recently, in some components of hydroelectric power stations. Modern oil rigs are designed to withstand up to 100 years of exposure to stresses caused by storms or hurricanes.

The coated electrode welding technique has the disadvantage of introducing defects and/or discontinuities in the base metal. According to ANDRADE (2010), porosity is a common defect in wet underwater welding. Research into atmospheric and wet welding shows that the high cooling rate leads to the formation of cracks in the thermally affected zone (HAZ) and in the weld metal (BRACARENSE *et al.*, 2010).

In general, the shipbuilding industry considers wet welding with a coated electrode to be a repair technique used in emergencies. For this purpose, commercial electrodes are used, such as AWS E6013 or electrodes with alterations to the composition of the rutile-based coating. (PESSOA, 2007).

This technique offers the best results when applied in shallow waters (depths of up to 20 metres), corresponding to the maximum depths of the structural components of floating units when in operation. However, it is necessary for the reliability of these repairs to be recognised by both the operators and the classification societies involved if structural integrity assessments are to be accepted from the point of view of quality for specific use.

Underwater welding usually refers to the AWS D3.6M:2010 "Underwater Welding Code" (AWS, 2010), where structural welds are classified into two categories. Class A welds have requirements for toughness, strength, ductility, hardness and bending at levels similar to those required by the main engineering codes in atmospheric welding, which are rarely achieved in wet welding. For class B welds, conceptualised as welds with limited structural quality, both the tests applied to qualify procedures and the acceptance criteria are less stringent than those for type A. Historically, wet welding procedures have been certified (qualified) as class B according to the international underwater welding standard AWS D3.6M. Another obstacle that makes it difficult to obtain class A, and is not directly related to the type of electrode, is the high hardness in the heat-affected zone (HAZ).

24 4 Main difficulties with wet welding

LIU (1999) states that the quality of wet welds is inferior to that obtained from dry welds due to various factors, such as the loss of alloying elements from the weld metal, increased hydrogen and oxygen content in the weld metal, which increases susceptibility to the appearance of pores and cracks.

Water is mainly responsible for the poor quality of wet welds compared to air welds, generating high porosity and low toughness (ISIKLAR and GIRGIN, 2011). The cooling rate in water is higher than in dry welding. In a wet environment, the cooling rate varies between 800 °C/s and 500 °C/s, while in a dry environment a rate between 415 °C/s and 55 °C/s is common (CHRISTENSEN, 1983).

In wet underwater conditions, the rapid cooling of the weld metal is the result of heat losses on the weld surface behind the electric arc.

Due to these factors, the main problems associated with wet welding are physical occurrences and metallurgical discontinuities, as described below (LIU et al, 1994): In the weld metal:
- Loss of alloying elements,
- Porosity,

- Cracking due to solidification.

- Hydrogen cracking,

In the ZTA:

-Excessive hardness

-Hydrogen cracking

Figure 2.2 shows a schematic of the main problems encountered in underwater welding, as described above.

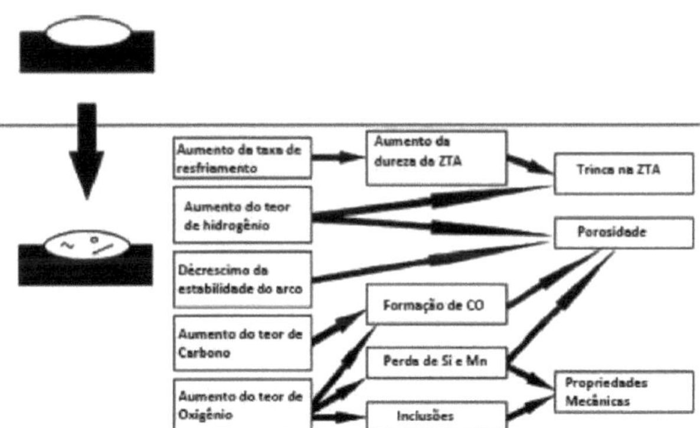

FIGURE 2.2 - Effect of water on wet welding.
SOURCE - Modified, LIU, 1994.

24.1 High hardness of the HAZ

The high hardnesses found in the HAZ of wet welds are the result of using steels with high hardenability, normally found in oil platform structures, in combination with the high cooling rates (BRACARENSE, 2008) found in wet welding. The cooling rate, in turn, depends on the heat input adopted.

In AH-36 structural steels used in class A approval attempts, maximum hardness values in the HAZ of around 375 HV were measured (PESSOA, 2013), i.e. above the limit permitted by the standard.

Hardness in the HAZ is mainly influenced by the chemical composition of the base metal and the cooling rate. This, in turn, is always very high due to the direct action of water, and is little influenced by the thermal input in wet welding with coated electrodes.

This is the situation found in the HAZ of the base metal caused by finishing passes. Therefore, in a conservative approach, the maximum hardness expected in the HAZ in C-Mn steels can be estimated by the well-known relationship between martensite hardness as a function of the steel's carbon content. Thus, considering the carbon and manganese contents specified for the main naval structural steels according to ASTM A 131, it is necessary to use special techniques to significantly reduce the cooling rate or reheat the

11

critical regions of the HAZ in order to meet the requirement for class A of AWS D3.MARINHO (2014) states that in tests carried out and published recently at LRSS- UFMG in partnership with PUC-RIO and CENPES, the maximum hardness in the heat-affected zone of the base metal of the finishing passes was higher than specified and did not meet the requirements of class A.

2.4.2 Cracks in the welding region

Cracks are considered one of the most serious types of discontinuities in welds. They develop when tensile stresses act on a weakened material, i.e. one that is unable to absorb these stresses through deformation. The types of crack that can develop are usually classified according to their location in the weld as: crater crack, transverse crack in the FZ, transverse crack in the HAZ, longitudinal crack in the FZ, margin crack, crack under the bead, crack in the bond zone and crack in the root of the weld. In ferritic carbon steels, hydrogen-induced cracks and reheat cracks can occur more frequently (ASME 1989).

2.4.2.1 Hydrogen-induced cold cracking

The main problem with wet welds is the high content of diffusible hydrogen, resulting from the dissociation of water due to the effect of the electric arc, which combined with high hardness increases the susceptibility to hydrogen cracking in the thermally affected zone.

YURIOKA, (1990) states that hydrogen cracking is caused by the simultaneous occurrence of the following factors: Sufficient concentration of diffusible hydrogen, microstructure susceptible to hydrogen, applied tensile stress of sufficient magnitude and temperature below approximately 200°C. These conditions are easily found in wet underwater welding, particularly in the HAZ (thermally affected zone) of the weld metal (KOU, 2003; SILVA, 1983; IBARRA et al., 1995; JOHNSON, 1997). In this region, the rapid cooling rate promotes the formation of martensitic microstructure in low carbon steels, making them susceptible to hydrogen cracking (GRUBBS, 1996).

The cooling rate, chemical composition and hardenability influence the microstructure of the weld bead. Wet welding conditions can form harder microconstituents in the thermally affected zone and in the weld metal, which are more susceptible to cold cracking (BAILEY 1994).

Stresses in the joint:

Residual stresses develop during the solidification of the metal and are influenced by thermal contraction and the degree of restriction of the joint. In contact with water there is a high thermal gradient, and the greater this gradient the greater the difference in thermal contraction, generating greater stresses. A greater degree of joint restriction raises the levels of residual stresses, increasing the likelihood of cracks occurring in the thermally affected zone and in the weld metal (BAILEY 1993).

The type of joint also influences restriction. A butt joint is less restricted than an angled joint. Stress concentrators play a very important role in hydrogen-induced cracking failure processes.

Presence of hydrogen:

GRANJON (1972) proposed a model for the flow of hydrogen during welding, as shown in Figure 2.3, where during solidification of the weld pool, the liquid is transformed into austenite, at which point part of the hydrogen is released into the atmosphere, the rest being retained in the austenite. As soon as this austenite decomposes into ferrite and cementite, the hydrogen diffuses into the austenite in the HAZ, due to the greater solubility of hydrogen in austenite.

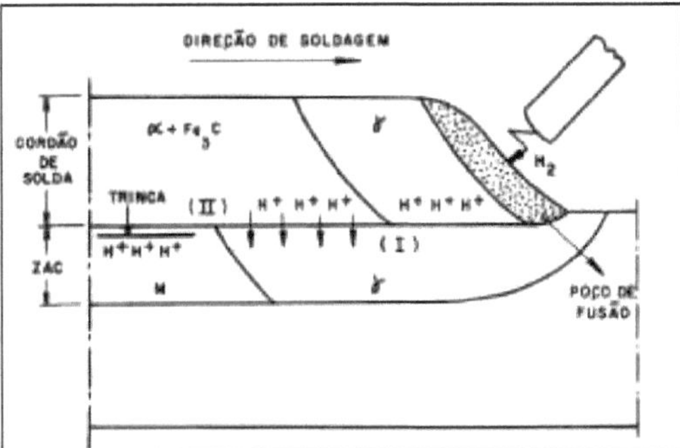

FIGURE 2.3 Schematic hydrogen flow in welding. SOURCE: GRANJON adapted, 1992.

Rutile electrodes have a high content of diffusible hydrogen, values in the order of 90ml/100g are recorded in the literature for metals deposited by this type of electrode in underwater welding (FILHO et. al, 2004).

Susceptible microstructure:

Hydrogen-induced cracks are favoured by the presence of microstructures with high hardness and low ductility, i.e. with a low capacity to accommodate the stresses resulting from the weld. Structures with low hardness can tolerate a greater amount of hydrogen without cracking. The hardness value of the microstructure can be used as a parameter to measure susceptibility to cracking. The hardness of the HAZ of a given steel depends on the cooling rate in the transformation range (usually 800 to 500 °C), its carbon content and its hardenability.

Wet welds in carbon steels with CE above 0.4% are more susceptible to hydrogen cracking in the HAZ (YURIOKA, 1982), because they tend to have high hardness structures in the HAZ, such as martensite.

The equivalent carbon, according to HW, can be calculated using the formula given in Equation 1 (CEnw):

$$CE = \%C + \%\frac{Mn}{6} + \%\frac{Cr+Mo+V}{5} + \%\frac{Ni+Cu}{15} \tag{1}$$

The four conditions necessary for hydrogen cracking to occur are met in wet welding as follows: High concentration of hydrogen in the weld joint due to the dissociation of water during welding. High hardness microstructure susceptible to cracking due to rapid cooling. High stresses generated by the high thermal gradient imposed on the joints. Finally, low temperatures (0°C to 30°C) which are the working temperatures of welded structures.

25 Sub-regions of the single pass welding region

In single pass welding, the weld is divided into three regions: the molten zone (MZ) consisting of the molten metal; the bond zone (BZ), consisting of a narrow region where partial fusion of the base metal occurs next to the molten zone and the heat-affected zone (HAZ), a region of the base metal that has undergone microstructural changes caused by the welding heat (BHADESHIA & HONEYCOMBE, 1980). The HAZ of steels can be further subdivided into the regions described below (BRANDI & WAINER, 1992), illustrated in Figure 2.8.

During welding, up to four sub-regions of the HAZ can be found, formed according to the maximum temperature reached and the time spent at this temperature. The formation of these sub-regions is determined by the characteristic transformation of the steel (WTIA, 2006). These sub-regions are listed below:

• Coarse grains: 1100 to 1500°C - liquid.

• Fine grain: 900 - 1030°C;

• Intercritical: 730 - 900°C;

• Subcritical: 500 - 700°C;

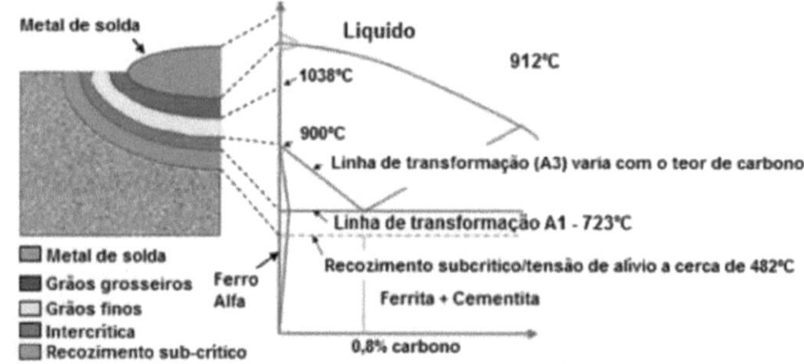

FIGURE 2.4 - Schematic of a single pass weld, showing the subdivisions of the HAZ.
SOURCE - SPERKO - 2006, p. 37.

• Coarse-grained ZTA (ZTA-GG) - this is the region adjacent to the fusion line (ZL), which is heated to temperatures between 1100 and 1500 °C, turning completely into austenite at the time of welding. Its final microstructure depends on the chemical composition of the material and the cooling speed. Temperatures close to the fusion line are well above A_C 3, which causes the austenite to overheat, resulting in a coarse-grained microstructure with high hardenability, which can have high hardness and low toughness, favouring the appearance of cracks.

- Fine-grained HAZ (HAZ-FG) - This region is heated between 900 and 1200 °C and is characterised by its smaller grain size. The further away from the fusion line, the smaller the grain size of the austenite, which reduces its hardenability and can cause its transformation into ferrite, producing a refined microstructure, with smaller grains than those of the base metal, corresponding to the typical microstructure of normalised steel and with good toughness.

- Intercritical HAZ (HAZ-IC) - HAZ region where the peak temperature range is between Aci and Ac3. In this range, the microstructure will be composed of austenite in the regions that have reached sufficient energy to undergo the transformation and untransformed ferrite and pearlite in lower energy regions. In this region, grain refining occurs at the end of cooling.

- Subcritical ZTA (ZTA-SC) - Found between 500 and 700 °C. As the peak temperature is lower than Aci, there are no visible structural changes. When welding hardenable steels, the heat input causes the microstructure to temper.

The peak temperature of the molten zone exceeds the melting point of the base metal. The chemical composition of the weld metal depends on the choice of consumables, the dilution rate of the base metal and the welding conditions.

The chemical composition of the HAZ remains unchanged over a wide range where the peak temperature has not reached the melting point of the base metal. However, there is a considerable change in the microstructure of the HAZ during welding due to the severe thermal cycle.

The metal immediately neighbouring the molten zone is heated within the austenitic field, where the precipitates formed in previous processes are generally dissolved. This region (ZTA-GG) is the one that presents the most problems in underwater welding, as it has excessively large grains, a factor that favours the formation of a very hard and brittle microstructure, martensite, and is therefore more susceptible to cold cracking.

2.6 Sub-regions of the multipass welding region

KÚCHLER (2009) says that multi-pass welding has some advantages over single-pass welding, which will be described below.

- Due to the thermal reheating cycles, each subsequent pass normalises and refines the grains of the previous layer.

- The thermal cycle causes tempering in the weld metal, reducing the residual stresses from the previous pass.

Figure 2.9 shows the schematic diagram of a multi-pass weld and the effect of overlapping tempering passes. It can be seen that part of the MS and ZTA are remelted by the subsequent pass.

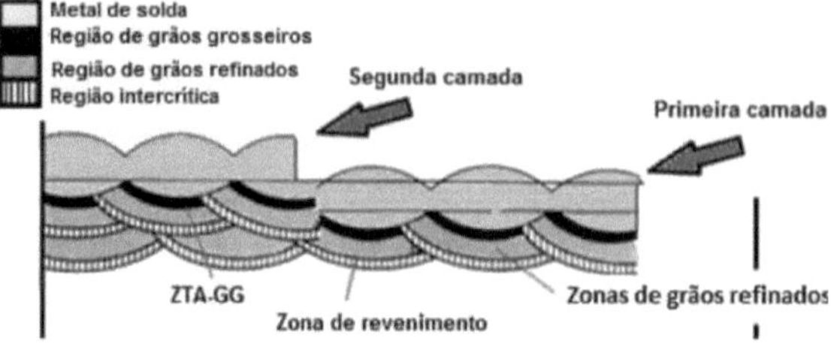

FIGURE 2.5- Schematic diagram of a multi-pass weld and the effect of tempering. SOURCE - ASM Metals Handbook, 1993.

2.7 Welding repair techniques without TTPS

When welded, hardenable steels have unhardened martensitic structures with high hardness and low toughness which, combined with residual stresses, can generate cracks. As a preventative measure, heat treatment is carried out to relieve internal stresses and improve mechanical properties. Welding these steels without further heat treatment is an attractive option, as heat treatments are generally expensive and can be very costly.

time-consuming due to the length of time they remain in the oven, and can also present difficulties when carried out on large structures.

Among the multipass welding repair techniques that aim to dispense with TTPS, the main ones are the half-layer and double-layer techniques which, through strictly specified and controlled welding procedures, take advantage of the heat transferred in the process itself to heat-treat the HAZ of the base metal in order to obtain a suitable microstructure with minimum toughness and maximum hardness requirements. These requirements are defined with the aim of guaranteeing the integrity of the repaired component, in particular preventing brittle fracture, reheat cracks or stress corrosion cracks.

2.7.1 Half-layer technique

The half-layer technique recommended by the ASME code aims to provide refining and tempering of the ZTA-GG of the first layer in a multi-pass weld, by overlapping the thermal cycles. The code specifies removing half of the first layer to facilitate overlapping the thermal cycle of the second layer. This technique is difficult to carry out due to the lack of control in removing half of the first layer by grinding and also because it is very labour-intensive.

A schematic of the half-layer technique is shown in Figure 2.10. This technique, adopted by the ASME Code, was developed for the repair of ferritic steel pressure vessels in nuclear power plants. As requirements, the ASME Code (1989) specifies:

• . Deposit the Iᵃ layer in the cavity to be repaired, using a small diameter electrode (maximum 2.5 mm).

• . Remove half the thickness of the Iᵃ layer by grinding.

• . Deposit the 2ᵃ layer with an electrode with a maximum diameter of 3.25 mm, and the subsequent layers with diameters between 3.25 and 4 mm.

• complete the joint with at least one layer of reinforcement, which will be removed so that the repaired surface is level with the rest of the part.

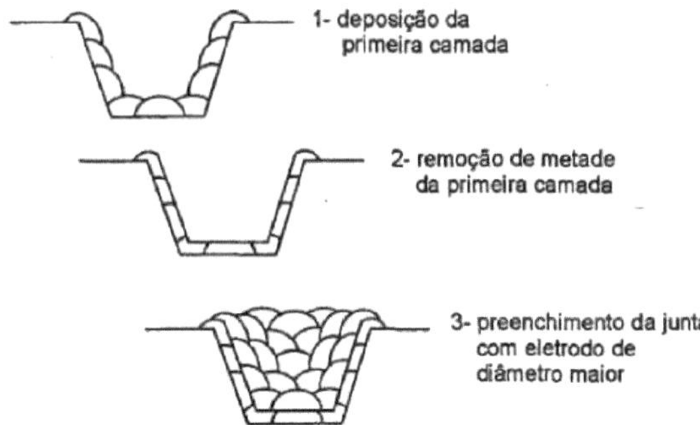

FIGURE 2.6- Schematic of the half-layer technique.
SOURCE - (HIGUCHI, 1980).

The applicability of this technique has been widely debated. HIGUCHI et al (1980) question the validity of the requirement to remove half of the I° layer, regardless of the thickness of the reinforcing layer which, for the same electrode diameter, can vary widely depending on the welding conditions.

2.7.   2Double layer technique

The coated electrode welding process is commonly used in double-layer repair operations because it is easy to apply in hard-to-reach places, can be used in all positions and does not require elaborate adjustments. However, as it is a manual process, it requires the welder's manual and technical skills.

In order to develop the double-layer technique, a number of actions, described below, are necessary (NINO, 2001).

-Estimate   the penetration, width and reinforcement of the weld bead depending on the welding conditions.

-Determine   the degree of tempering and grain refining through welding thermal cycles.

-Estimate   welding thermal cycles as accurately as possible and determine the distribution of peak temperatures in the workpiece.

Low hydrogen consumables should be used to minimise the risk of cold cracks and preheating should be used to control cooling, which facilitates hydrogen diffusion and reduces residual stresses.The double layer technique, as described above, consists of promoting a suitable overlap of thermal cycles in such a way that the second layer promotes refining and tempering of the ZTA-GG of the first layer, as shown in Figure 2.11. The literature suggests that the energy of the second layer must be greater than that of the first to cause refining in the ZTA-GG of the first layer.[a] . This ratio must be determined according to the welding speed, the type of electrode and the parameters of the welding process adopted (NINO, 2001).

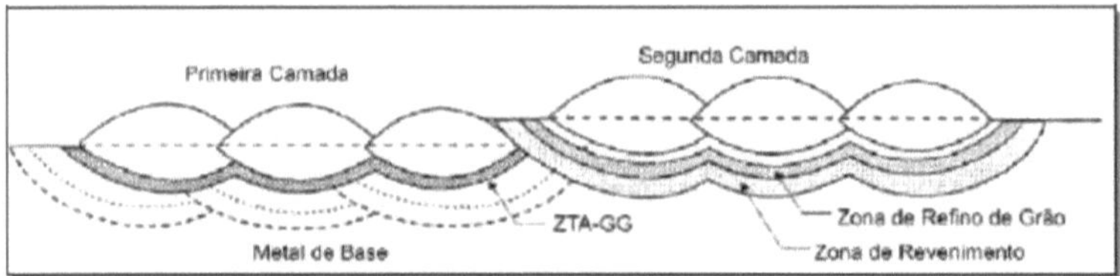

FIGURE 2.7- Diagram of the sub-regions of the HAZ in a multi-pass weld.
SOURCE - NINO, 2001.

Welding processes used to apply the double layer

The coated electrode process is commonly used, as it can be easily applied in hard-to-reach places, in all positions and does not require sophisticated adjustments to the process conditions (TEIXEIRA, 1992). However, as it is a manual process, the results depend largely on the welder's manual skill and technical knowledge of the process itself and the objectives sought with the application of the double layer technique.

2.8 Tempering pass

This welding technique was developed specifically to refine the microstructure in the metal by carefully depositing weld beads and controlling the welding energy. This heat treatment aims to improve fracture toughness and reduce the high hardness of the HAZ; (WTIA, 2006).

The tempering pass technique was used by AZEVEDO (2002) who applied the double layer technique to AISI 1045 steels, where he achieved the following toughness results

similar to the specimens heat-treated after welding, using various energy ratios.

23.1 Application of the tempering pass

The treatment of the weld bead by a tempering pass was developed to reduce the hardness of the MS and HAZ. There are claims (NINO, 2001) that the mechanical properties of the HAZ can be improved similarly to the improvements obtained with conventional heat treatment.

The tempering pass aims to reduce the hardness of the HAZ by imposing its thermal cycle. To do this, the tempering pass must be positioned appropriately in relation to the pass to be tempered, so that the ACi isotherm of the subsequent layer touches the fusion line of the previous layer, so that the base metal is reaustenitised and the rest of the HAZ is tempered below the transformation temperature range (WTIA, 2006). Figure 2.12 shows a schematic of the softening of the HAZ with a tempering pass.

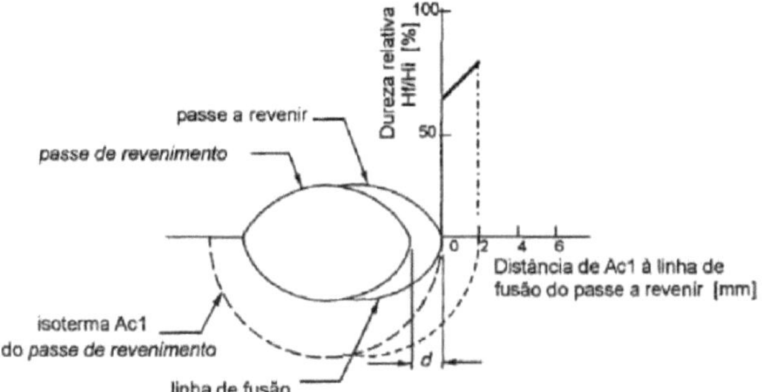

FIGURE 2.8- Relative drop in hardness as a function of the distance between the LZ.
SOURCE - (WTIA, 2006).

OSLEN et al (1982) further describes the tempering pass as a technique widely used to reduce the hardness of the HAZ, with the aim of reducing the chances of hydrogen-induced cracks occurring. The tempering pass, when specified, would be treated as an essential variable (i.e. if it is used in the qualification of a welding procedure, it should also be used in field welding).

However, it is very difficult to implement this technique in the field due to the high degree of difficulty in controlling the bead deposited by welders when subjected to the underwater environment.

2.8.2 Martensite and tempering

The transformation of austenite into martensite occurs through a rapid shearing process, which is not normally accompanied by diffusion. As a result, a supersaturated solid solution of carbon in iron is obtained, which has a tetragonal body-centred structure, which is a distorted form of CCC iron (SHEWMON, 1969).Figure 2.13 shows the influence of carbon content on the hardness of martensite, as well as the large difference in hardness between martensite, pearlite and spheroidite. The hardness of martensite is independent of the content of substitutional alloying elements, which are added mainly because of their effect in delaying the formation of pearlite and bainite.

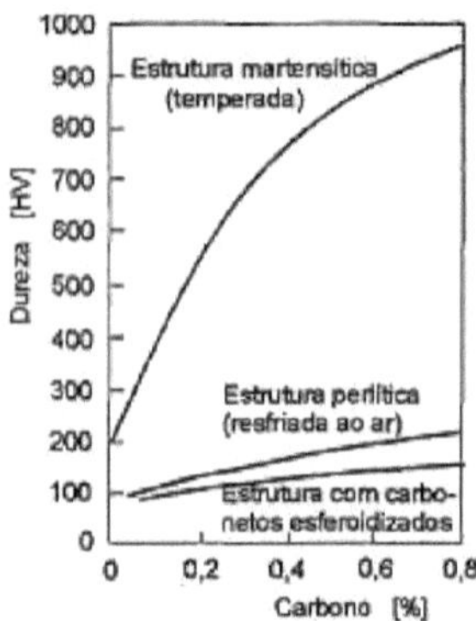

FIGURE 2.9- Influence of carbon content on the hardness of martensite, pearlite and a structure of spheroidised carbides. SOURCE - SHEWMON, 1969.

According to CHIAVERINI (1977), tempering consists of an increase in temperature below that of the critical zone and is applied to hardened steels, resulting in modifications to the structure obtained during hardening. There is an improvement in ductility and a reduction in hardness at the same time as internal tensions are relieved or eliminated. The final properties of the tempered material depend on the time and temperature of the process, and the structural modification can be intense, resulting in better machinability conditions (lower hardness values). Table 2.2 (in topic 2.10.1) shows the hardness and elongation values for ABNT 1045 steel at different tempering temperatures.

2.9   - Some of the elements needed to develop the double layer technique

Some of the elements needed to be able to model this technique are:
- Know how the material responds to welding thermal cycles in terms of tempering, grain refining, precipitation and segregation phenomena.
- Estimate the geometry of the weld bead (penetration, width and reinforcement) as a function of the welding conditions, including the type of process, the values of current, voltage and welding speed, the welding position and the characteristics of the consumables.
- Being able to calculate the welding thermal cycles and the temperature field, i.e. the distribution of peak temperatures in the workpiece.

The main parameters to be observed for a successful operation are described below
- Dimensions of the first layer bead.
- Appropriate ratio between the welding energies of the layers.
- Preheat and interpass temperature control during welding.

2.10 ABNT 1045 steel

According to NBR 172/2000, ABNT 1045 steel is classified as carbon and special construction steel (ABTN, 2000).

According to standard NBR NM 87/2000, which establishes the chemical composition of steels for mechanical construction, ABNT 1045 steel must have the chemical composition described in Table 2.1.

TABLE 2.1- Chemical composition of ABNT 1045 steel (% by mass).

C	Mn	P maximum	S maximum	Yes
0,43-0,50	0,60-0,90	0,04	0,05	0,10-0,60

SOURCE - ABTN, 2000.

2.10.1 Applications and mechanical properties of ABNT 1045 steel

ABNT 1045 steel is used, among other things, in distillation tower pump shafts and fan shafts used in oil and gas refineries, which generally suffer wear and tear due to the aggressive environment to which they are exposed.

Because ABNT 1045 steel contains between 0.43 and 0.50 per cent carbon, when subjected to thermal welding cycles it develops high hardness in the heat-affected zone (HAZ) and has low resistance to tempering. Table 2.2 shows the mechanical properties of ABNT 1045 steel as a function of tempering temperature.

TABLE 2.2- Mechanical properties of ABNT 1045 steel.

Tempering temperature (°C)	Elongation (%)	Brinell hardness (HB)
205	09	363
260	11	352
315	09	341
370	15	331
425	08	311
480	18	293
540	20	277
595	22	255
650	24	241

SOURCE - METALS HANDBOOK, 1991.

2.102 Temperability of ABNT 1045 steel

Hardenability has been used as an indicator of weldability and as a guide for selecting materials and processes in order to avoid excessive hardening and, consequently, the occurrence of cracks in the HAZ. Steels with high hardenability provide a high volume fraction of martensite in the ZF and ZTA. As is well known, depending on their carbon content, this microstructure can be highly susceptible to hydrogen-induced cracking (BAILEY, 1994).

Various empirical equations have been developed to express the weldability of steels. Carbon equivalent (CE) equations were the first expressions used to estimate the susceptibility of steel to cracking in the welding process and also to determine the need for preheating and post-heating to prevent cracking. These equations take into account the effects of various chemical elements on the hardenability of steel.

The GRAVILLE diagram (1976) shown in Figure 2.14 shows the susceptibility to hydrogen-induced cracking as a function of carbon content and equivalent carbon. In this figure, zone I is typical of low-carbon steels with relatively low hardenability, which are not very susceptible to hydrogen-induced cracking. In zone II, the steels have higher carbon than those in zone I, but low hardenability, so it is possible to avoid crack-sensitive microstructures by controlling the cooling of the HAZ. This can be achieved by controlling the welding energy and using preheating. In zone III, steels have high carbon contents and high hardenability, which can easily produce crack-sensitive microstructures. To avoid hydrogen-induced cracking in this zone, low hydrogen processes, preheating and post-weld heat treatment should be used (METALS HANDBOOK, 1992).

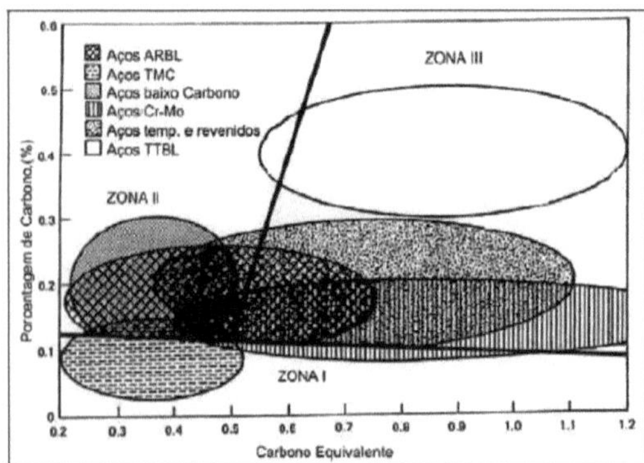

FIGURE 2.10- Graville diagram.
SOURCE - METALS HANDBOOK, 1992.

Figure 2.14 shows that steels with a medium carbon content fall into zone II and therefore require control of the cooling of the HAZ when welded. When dealing with thick steels, certain precautions are necessary because the higher cooling speed can lead to the formation of fragile microstructures (METALS HANDBOOK, 1992).

The structure of the coarse-grained HAZ of a given steel can be predicted using Continuous Cooling Transformation (CRT) diagrams developed especially for welding. They are similar to the usual TRC diagrams, except that

its high austenitisation temperature of around 1300°C. The TRC diagram for the HAZ has characteristics that affect the austenitic grain size and its homogeneity, making it difficult to apply it to predict the HAZ microstructure of a real weld (MODENESI, 1992).

Figure 2.15 shows the TRC diagram for ABNT 1045 steel. It can be seen that close to the martensite transformation start temperature, Mi, for the same cooling rate, the welding transformation curve shifts to the right in relation to the heat treatment transformation curve, so the probability of martensite forming in this steel is greater during welding than during heat treatment. An example, shown in Table 2.3, with data taken from Figure 2.15, indicates that for a cooling rate of 30° C/s, 92% martensite can be obtained during welding and only 69% during heat treatment. This may be related to the grain growth of austenite that can occur during welding, reducing the grain boundary area available for the nucleation of intermediate constituents during cooling (KOU, 1987).

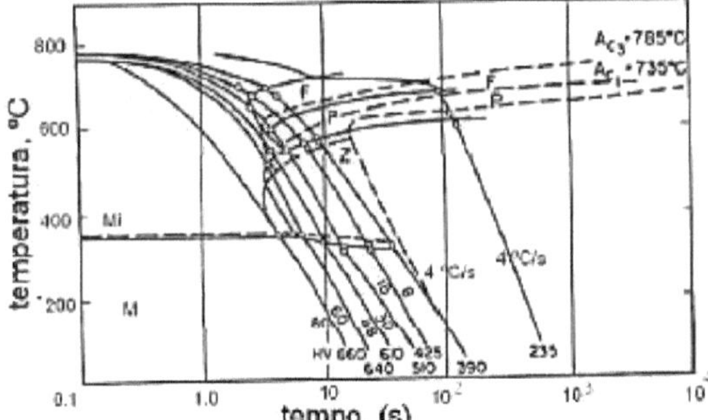

FIGURE 2.11- TRC diagrams of an ABNT 1045 steel during heat treatment (dotted lines) and during welding (solid lines).

SOURCE - KOU, 1987.

TABLE 2.3- Microstructure of ABNT 1045 steel due to welding and heat treatment

Cooling rate CC7s)	Microstructure (%)		
	Ferrite	Periite+intermediate structure	Martensite
4	5(10)[a]	95(90)	0(0)
10	1930	9(70)	90(27)
30	1(1)	7(30)	92(69)
60	0(0)	2(2)	98(98)

a- The values between the' and the dates refer to the Dercertua. de tase for the term treatment.

SOURCE - (KOU, 1987).

Grain contours are a favoured location for the nucleation of new phases. The larger the size of the austenitic grain, the smaller the number of contours per unit volume, therefore the longer the austenite incubation time and the higher the steel's hardenability. This fact justifies the ease with which the ZTA-GG forms martensite.

Since the maximum temperature reached and the cooling rate vary from point to point in the HAZ and each cooling curve crosses the diagram at specific locations, the HAZ has different microstructures at each point. Close to the fusion line, when welding C-Mn and low alloy steels, the likely microstructures are martensite and bainite for a suitable carbon equivalent (AGUIAR, 2001).

2.11 Hardness test

Perhaps because hardness is an easy characteristic to measure (either in the laboratory or in the field), numerous studies have sought to correlate it with the microstructures and properties of materials. In welding, this characteristic is even more valuable, given the multiplicity of microstructures present in the HAZ, due to the high temperature gradients. For example, today hardness limits of 350 and 248 HV are specified in the literature and in welding manufacturing standards, respectively, to avoid cold cracks (COE, 1973) and stress corrosion cracks (NACE, 1990).Figure 2.16 shows a schematic of the Vickers hardness measurement, where in A the penetrator is making the indentation and in B the indentation mark produced by the penetration. It can be seen that the lower the hardness of the material, the lower the material's ability to resist penetration and, consequently, the higher the dl and d2 values.

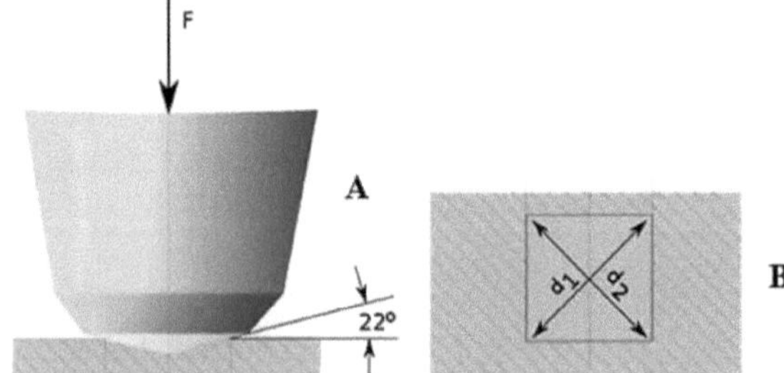

FIGURE 2.12 - Vickers hardness print measurement.
SOURCE: WIKIMEDIA, 2017.

CHAPTER 3

METHODOLOGY

The main objective of this work is to evaluate the effect of grinding and direction reversal on the microstructural properties and consequently on the mechanical properties of the Thermally Affected Zones (HAZs) of joints welded using the wet underwater welding process. It is therefore important to know the base material used, the consumable electrode that will carry out the welding and the welding energy imposed by the process. This chapter describes how the choice of each material and welding parameter used was made, also presents the methodology used in the grinding and direction reversal processes when these were employed and also how the properties were evaluated in order to check whether or not there was an improvement in the properties of the welded specimen.

3.1 Materials

3.1.1 Base metal

In order to guarantee the necessary chemical composition for the HAZ to have higher hardness values, we opted to use SAE 1045 carbon steel, with the chemical composition shown in Table 3.1 below.

TABLE 3.1 - Chemical composition of the ABNT 1045 steel used (% by mass).

C	Mn	P	S	Yes
0,46	0,68	0,022	0,02	0,17

In addition to the chemical analysis of the material provided by the supplier, metallographic analyses were carried out to characterise the material. Figure 3.1 shows the microstructure of the base metal used as received at 200X magnification.

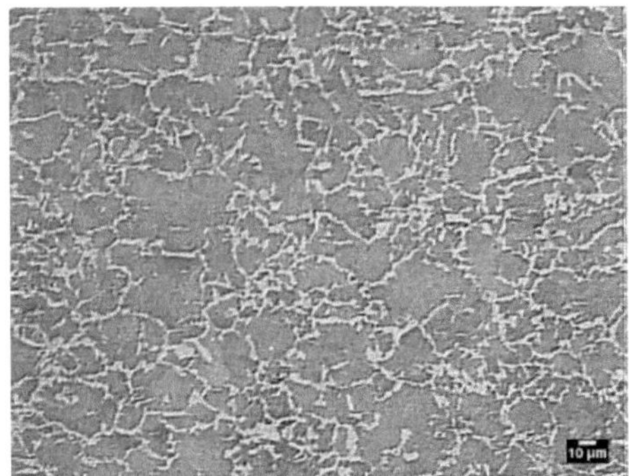

FIGURE 3.1 - SAE 1045 steel used, 200X magnification.

SOURCE - Author.

Still with the intention of characterising the material and checking whether it would meet the purpose of the work, which is to obtain high hardness values when subjected to sudden cooling, a 40x40x40[mm] specimen was heated in an oven to a temperature of 900°C for 30 minutes and then tempered in water. After hardening, surface hardness tests were carried out and the surface was analysed metallographically. The hardness reached values of around 600 HV5 (load applied for 15 seconds).

3.1.2 Chamfer preparation

The main purpose of the chamfer configuration used in this work was to generate an HAZ with a well-behaved geometry, i.e. with this configuration it is easier to predict the extent and location of the HAZ as well as each of its regions (hard zone and softened zone).

Figure 3.2 shows the chamfer that was developed especially for this job. It is 200mm long, ensuring that the entire weld bead is deposited within the chamfer. In addition, the central parts of the bead can be used for the analyses, so there is no need to use regions of instability such as the arc opening region.

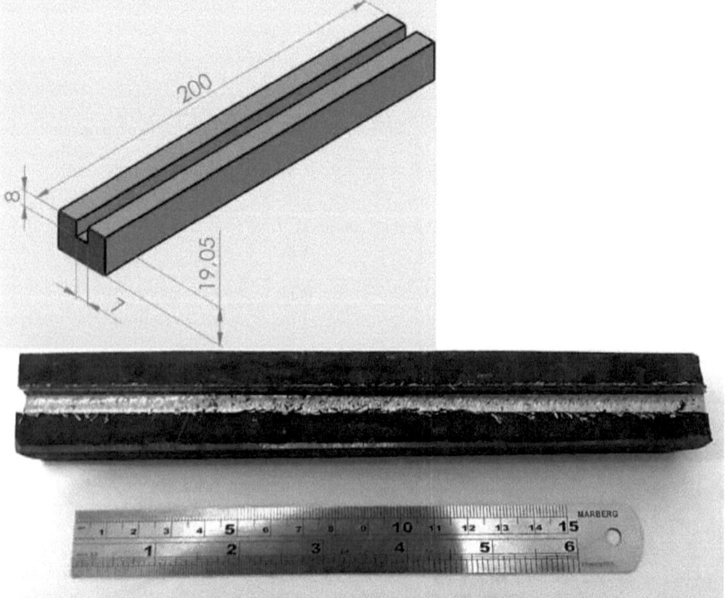

FIGURE 3.2 Parallel edge chamfer used in this work. SOURCE - Author.

Figure 3.3 shows the cross-section of the chamfer used for filling. This was developed to understand the development of the HAZ with each pass. It has an opening of 7mm and its walls are parallel to each other.

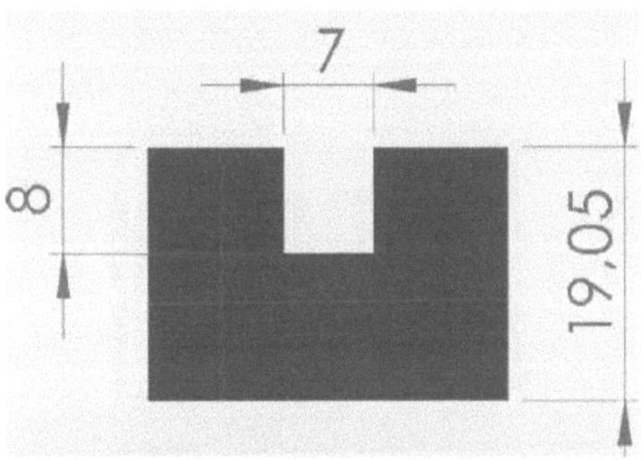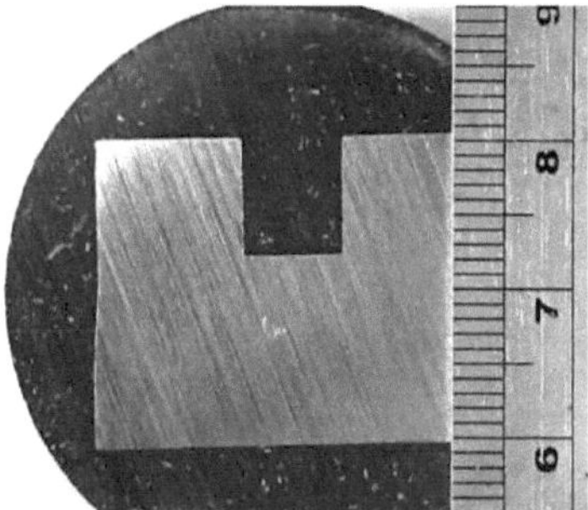

FIGURE 3.3 - Parallel edge chamfer used in this work.
SOURCE - Author.

3.13 Electrodes

The electrodes used in this work were of the AWS E 6013 rutile type and conformed to the AWS standard, with a nominal gauge of 3.25mm.

Varnishing the electrodes

To prevent the electrode coating from absorbing water when it is submerged, all the electrodes received two layers of vinyl varnish. This layer is waterproof and prevents water from degrading the properties of the coating, making welding possible. The colourless varnish used is suitable for this application, with a formulation based on vinyl ketone resin.

Each electrode received a layer of varnish and was placed on a support to drain off the excess product and dry the first layer. A second layer of varnish was applied 24 hours later and the electrodes were again left on the support to drain off the excess varnish and dry.

3.   2Welding equipment

The equipment used to carry out the work was made available by the LRSS - Welding and Simulation Laboratory of the Federal University of Minas Gerais, ranging from machines dedicated to carrying out underwater welding to equipment for cutting and analysing metallographic parts.

32.1 Hyperbaric tank

To simulate the desired depth of 10 metres, a hyperbaric tank was used, capable of simulating depths of around 150m. The 10 metre depth was chosen on the basis of work carried out at the LRSS itself, where this depth has a lower occurrence of porosity in the weld metal. The depth was therefore the same as that used for all the test specimens. Figure 3.4 shows the tank used for this work. To simulate the depth, compressed air is blown in to the desired pressure. This can be controlled and checked by a pressure gauge installed in the tank itself.

FIGURE 3.4 - Hyperbaric tank used to make wet welds and simulate depth.

3.2.2 Welding source

The power source used to weld all the specimens was a machine dedicated to carrying out wet underwater welding and operates with characteristic curves and dynamic conditions that improve the stability of the process underwater. The IMC Hiper-1 underwater power source is shown in Figure 3.5.

FIGURE 3.5 - IMC Hiper-1 welding source used for underwater welding of the specimens.

323 Gravity welding system

The gravity welding system for SMAW is based on the inclination and melting rate of the electrode, and the welding speed is based on the fusion of these two variables. Various tests were carried out to find the best welding conditions, mainly assessing visual aspects such as the absence of bites, excessive reinforcement and surface pores. Figure 3.6 shows the diagram with the important angles that influence the welding speed of the process.

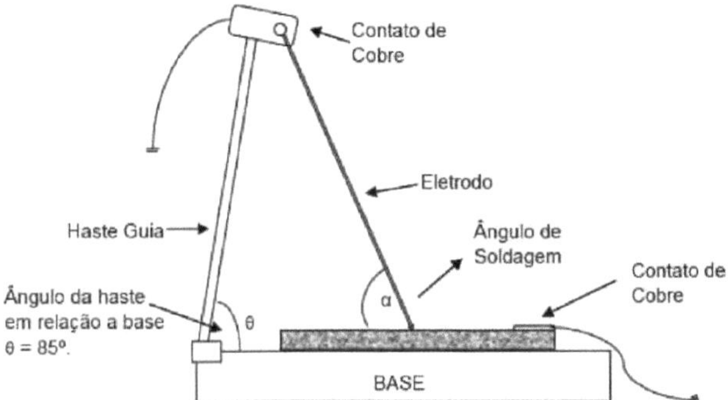

FIGURE 3.6- Schematic of the mechanised gravity welding device used in this work.

The system has two angles a (60°) and 0 (85°). Once the angles were selected, they were no longer changed, thus avoiding the influence of different welding speeds on the specimens.

33 Welding energy calculation

As shown, welding energy is extremely important in the extent and behaviour of the HAZ, and this zone is the main objective of the research. MODENESI (2003) states that welding energy (H) can be calculated as shown in Equation 3:

$$H = \frac{VI}{v}$$

(3)

Where:

V   - Average process voltage

I - Average process current

V   - Average welding speed

The average welding speed was obtained by dividing the length of the deposited bead (Z) by the open arc time (t) during the process. Equation 4 below shows the formula for calculating the average welding speed.

$$v = \frac{l}{t} \tag{4}$$

33.1 Current and voltage acquisition system

For greater certainty in calculating the welding energy of the process, the SAP V4 acquisition system was used, capable of monitoring and recording the current and voltage values at the time of welding. The acquisition rate was 5,000 points per second. Figure 3.7 shows the SPA V4 equipment used.

FIGURA 3.7-   SAP V4 system used to acquire process current and voltage.

With the help of this system, it is possible to calculate the heat input for each welding situation, and thus know it when the best welding parameters are set. Figure 3.8 shows an oscillogram captured using this system.

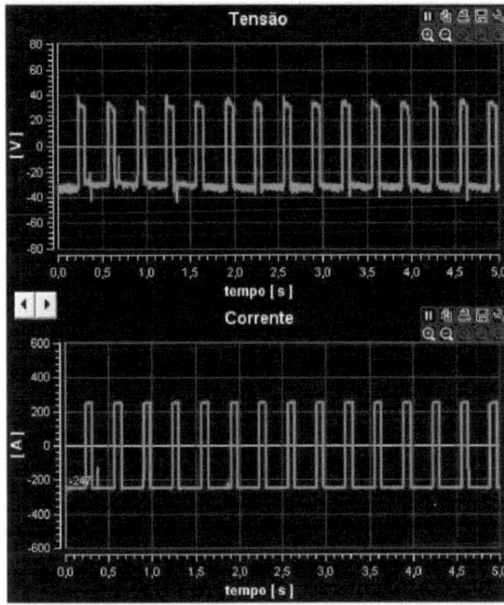

FIGURA 3.8-   SAP V4 interface - Signal acquisition system used.

SOURCE-http://www. imc-soldering.com.br/en-br/equipamentos/sap-v4

3.4 Grinding method

One of the main variables in this work to study the improvement of the mechanical properties of submerged welded joints is based on the thinning of the reinforcement of each weld bead immediately after its deposition, this thinning was produced by controlled grinding, always leaving 1 mm of reinforcement with each welding pass made.

Grinding was carried out with an Imm abrasive disc and abundant cooling. Height control was carried out with a caliper, which had the surface of the bevel as a reference point, then the depth to the top of the bead was measured, always maintaining the height Imm with each welding pass. Figure 3.9 shows the result of grinding and how the height control was carried out.

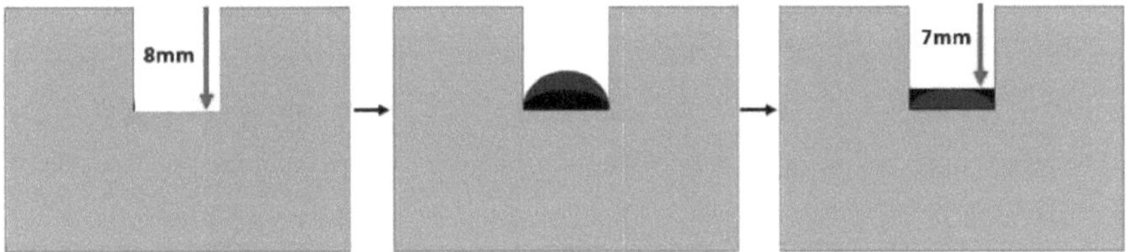

FIGURE 3.9- Schematic of the cross-section of a deposited strand and the cross-section of the same strand after the reinforcement grinding process.

33 Reverse direction method

The other technique studied to see whether or not it improves mechanical properties is the method of

31

reversing the welding direction between subsequent passes. With this method it is hoped to obtain a difference in the direction of growth of the columnar grains. This phenomenon occurs because the columnar grains present in the centre of the weld bead grow following isotherms, which in turn have a direction in line with the welding direction. Figure 3.10 exemplifies this method.

DIREÇÃO DE SOLDAGEM

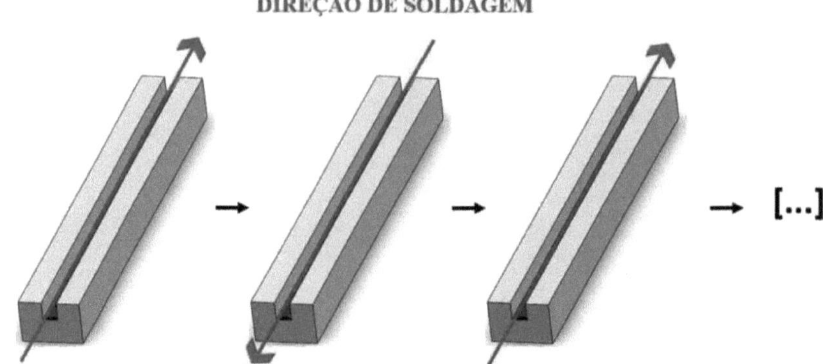

FIGURE 3.10- Method of reversing the welding direction.

Reversing the welding direction basically varies the solidification angle of the weld metal's columnar grains. When the columnar grains of the subsequent pass have their direction of growth altered, the grain orientation of the weld metal presents a longitudinal "zigzag" that hinders the propagation of cracks and tends to increase the toughness and mechanical strength of the weld metal.

The aim of applying the welding direction inversion technique to each pass is to change the orientation of the columnar grains, resulting in better weld metal properties. The application of this technique is also intended to verify whether the change in solidification direction affects the thermal cycle between subsequent passes in the HAZ. As explained in section 2.4.4, the speed of solidification in the underwater process differs from the dry process, so the application of the inversion technique aims to explore a possible change in heat flow due to the change in welding direction, which would reflect possible changes in the formation and mechanical properties of the HAZ.

3.6 Cutting the cross-section per pass

The main way of analysing the characteristics of the welded joint and its mechanical properties in this work is through hardness testing and metallographic analysis of the welded region. In order to study how the joint behaves with each pass, the methodology developed in this work is to cut out a section of the specimen with each weld bead made in the chamfer. Figure 3.11 shows how the section to be analysed is removed from the specimen.

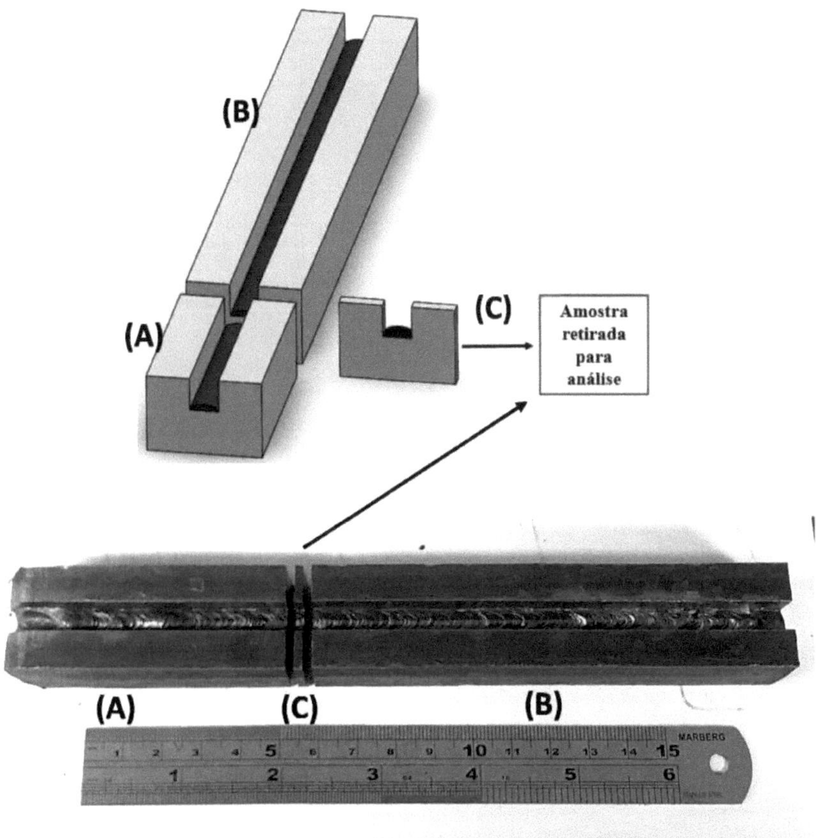

FIGURE 3.11- Cut specimen with the sample removed highlighted.

After making a weld pass and then removing a section (C), the remaining parts of the specimen (A and B) were prepared again and repositioned in the original direction. Care was taken to keep the faces parallel to each other, eliminating any possible spacing between the sectioned specimens. Figure 3.12 shows this process of repositioning the specimen.

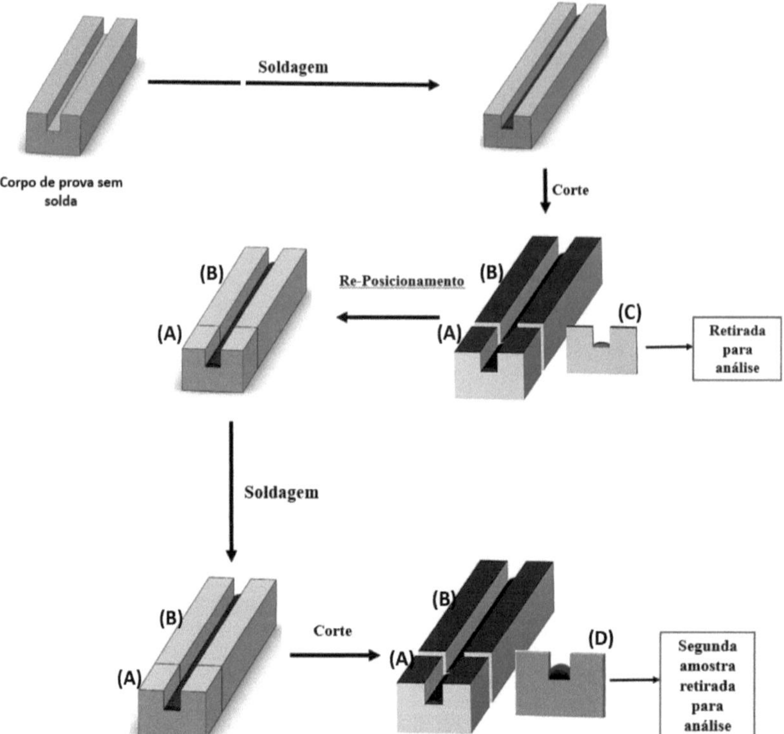

FIGURE 3.12 - Removal of the second cross-section (D) of the welded joint. At this point the section has 2 beads deposited.

With the specimens (A and B) already repositioned, a weld pass is again made on the chamfer. After welding, a cross-section of the joint (D) is again removed, immediately adjacent to the area that was previously removed (C).

With the grinding process

In the case of the specimens that will be used in the grinding process, a new stage is inserted into the joint manufacturing cycle. The specimen undergoes a grinding process, leaving only 1 mm of reinforcement of the deposited bead, as already discussed in the methodology, and only then after the grinding process are the specimens repositioned to carry out a new welding pass.Figure 3.13 below shows the manufacturing scheme for a specimen that undergoes grinding.

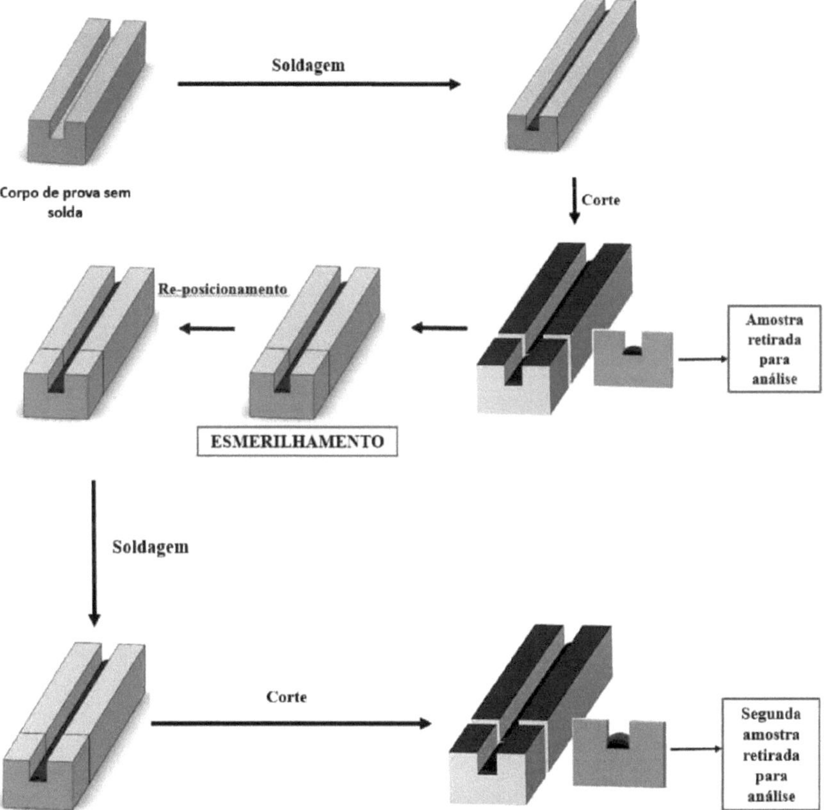

FIGURA 3.13-   Step-by-step diagram for welding the chamfer when using the grinding process.

With the process of reversing meaning

In the case of specimens to be used in the direction reversal process, the specimen is rotated during repositioning so that the welding direction is the opposite of the direction of the immediately preceding pass, meaning that the end of the bead in one pass will be the beginning of the bead in the subsequent pass.

Figure 3.14 below shows the manufacturing diagram for a test specimen that undergoes the welding direction inversion process.

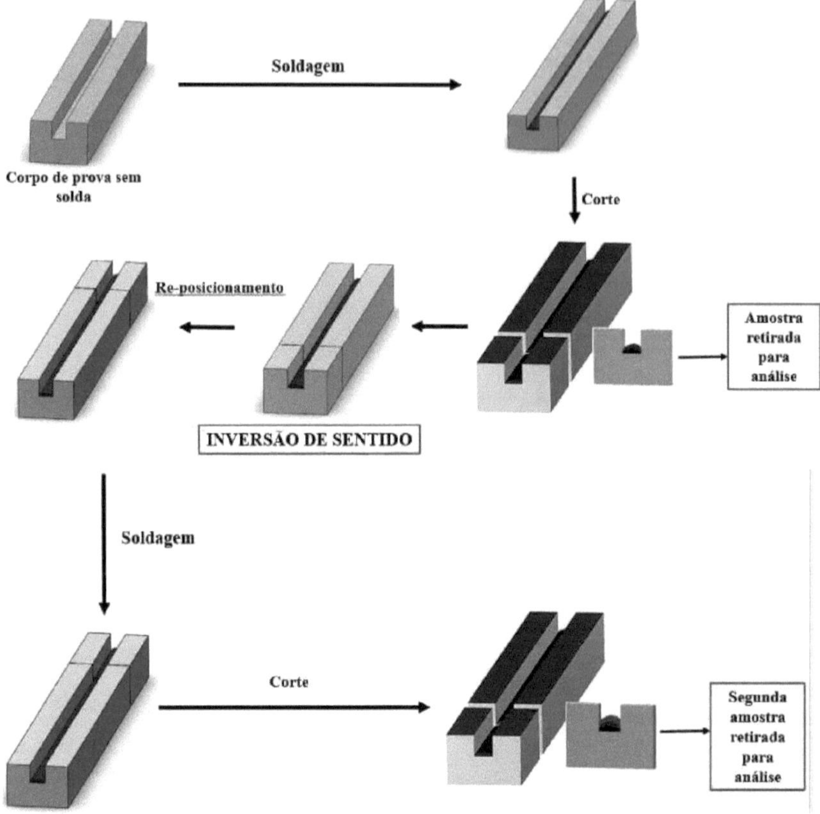

FIGURA 3.14-   Diagram of the step-by-step process for welding the chamfer when using the reverse welding direction process.

Figure 3.15 shows an example of the same specimen welded in this work using the direction reversal method. Note the variation in the direction of the "scales" produced by the bead. This example is of a specimen that had already been sectioned for sample removal, as mentioned earlier in the paper.

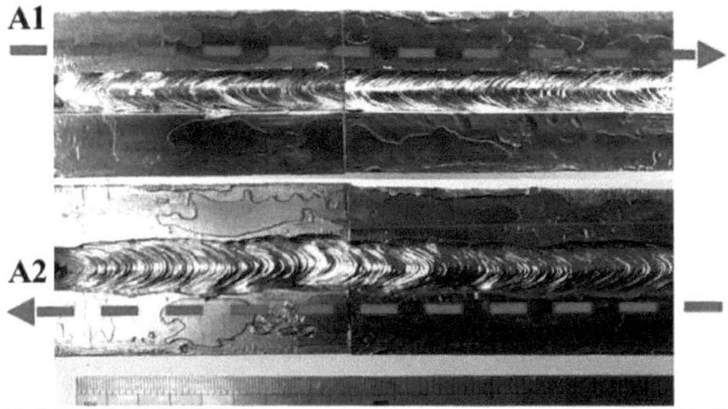

FIGURE 3.15- The same chamfer welded with the reverse direction process. The figure shows the same specimen in subsequent beads (Al) and (A2).

SOURCE - Author.

35.1 Board history

By taking a cross-section of the welded area for each pass, a kind of "Joint History" is obtained, where the mechanical behaviour of the same point in the HAZ will be studied for each pass made. This behaviour will be evaluated using the microhardness test, which will be explained later. At the end of welding the chamfer, there are several "slices" of the welded area, as shown in Figure 3.16.

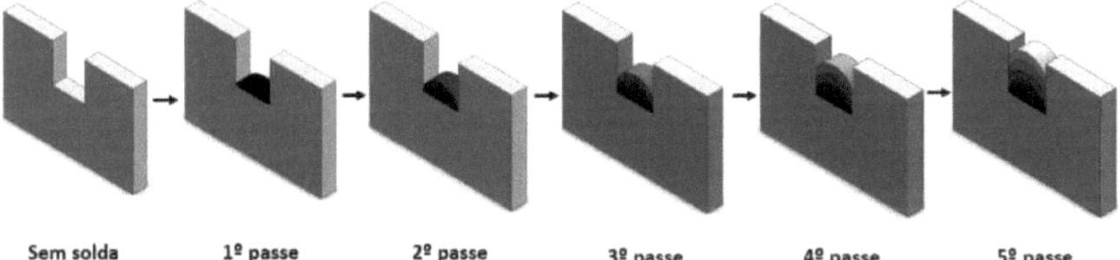

FIGURE 3.16 - Diagram of the specimen cross-sections.

1.1.1 Positioner

In order for the procedure shown above to be carried out properly, it is necessary to ensure close positioning between the parts to be welded, for which a device has been developed that prevents the parts from moving. The parts are placed on the device shown in Figure 3.17 and then locked. The device with the parts already assembled is placed in the tank for each welding pass.

FIGURE 3.17- Dedicated positioner for the specimen used in this work. It ensures good positioning and avoids gaps between the parts to be welded.

3.7 Vickers microhardness test

The microindentation tests were carried out on a SHIMADZU HMV Digital Microindenter, with a pyramidal diamond penetrator with a 136° taper. The measurements were taken by moving the specimen on the bench of the device, with observation through a 40x magnifying lens. A load of 4.903 N was applied for 15 seconds. Figure 3.18 shows the machine used in this test.

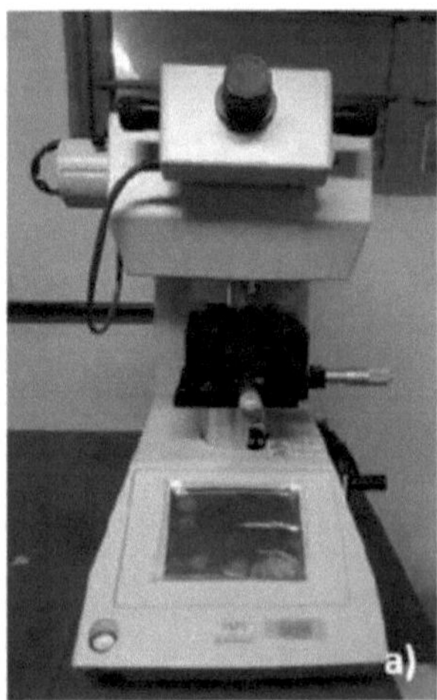

FIGURE 3.18- Digital micro-identifier used in the tests.

To analyse the hardness profile of the welded region, especially the HAZ, a methodology was developed that was repeated in all the samples analysed. Forty-two identifications were made on each sample, 21 on each side of the joint and always in the regions identified as the HAZ. The identifications were made

0.5mm apart and 0.5mm away from the Bond Zone, where the HAZ begins. Figure 3.19 identifies these regions in the welded joints.

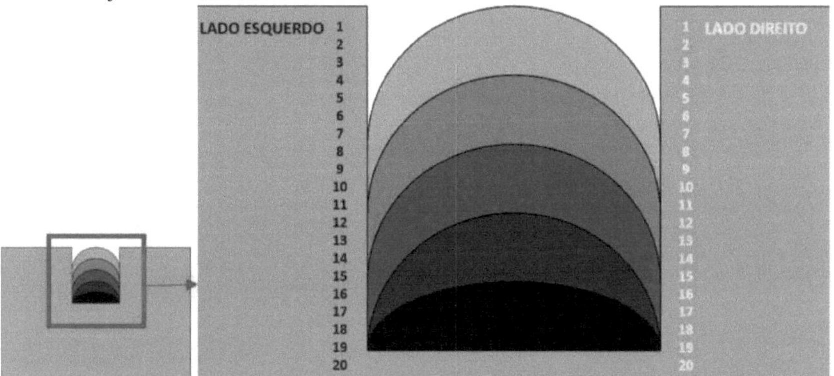

FIGURE 3.19- Schematic of a chamfer filled with 5 passes, highlighting the regions where the microhardness measurements were taken.

This profile aims to quantify the hardness values along the HAZ and identify, above all, whether there were regions where tempering was not effective and sufficient to promote softening of the material.

3.8 Metallography

Metallographs were taken with the main aim of evaluating the resulting microstructure and qualifying the ecrystallised region after the chamfer had been completely filled.

In order to analyse the joint as a whole, a methodology was developed to obtain an image of the entire welded region from localised images that allow the microstructure to be assessed. In other words, the macrograph used to assess the tempered region was obtained from several micrographs. The micrographs were taken on the Olympus SZ- CTV and Olympus BX60M mircroscope, with photometer and image capture camera, shown in Figure 3.20. The images were captured using HLImage software.

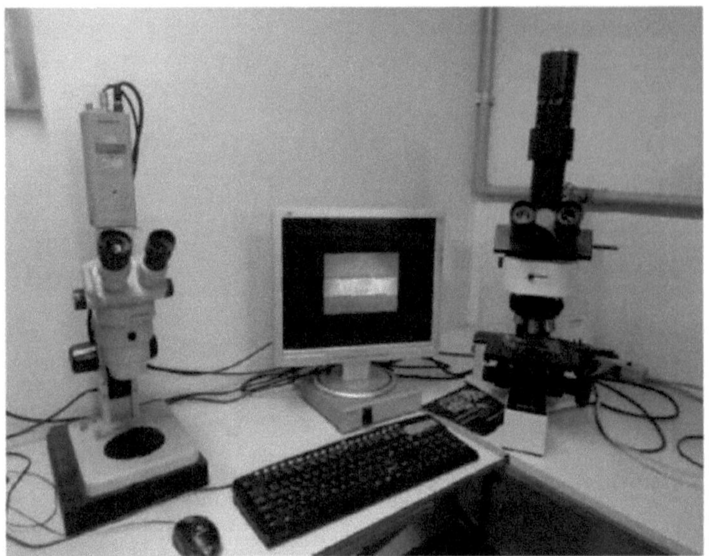
FIGURE 3.20 - Olympus SZ-CTV and Olympus BX60M microscopes used for micrographs and macrographs.

3.8.1Macrographs made by mounting micrographs

To assemble the micrographs, which have a magnification of 25x, Adobe Photoshop CS6 was used. Figure 3.21 shows a comparison between a macrograph taken conventionally, from a single photo, and the macrograph obtained using the methodology used in this work.

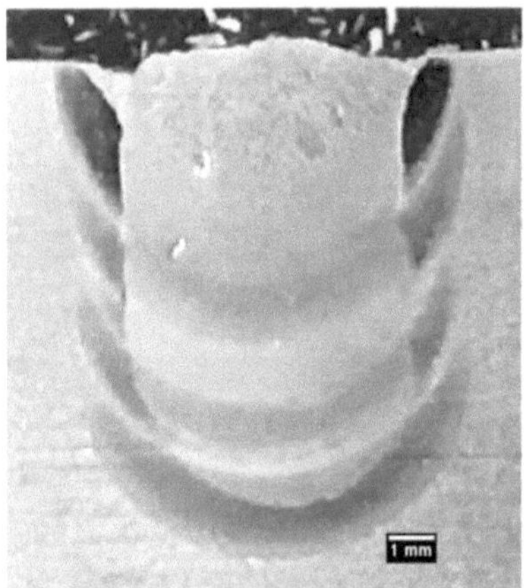

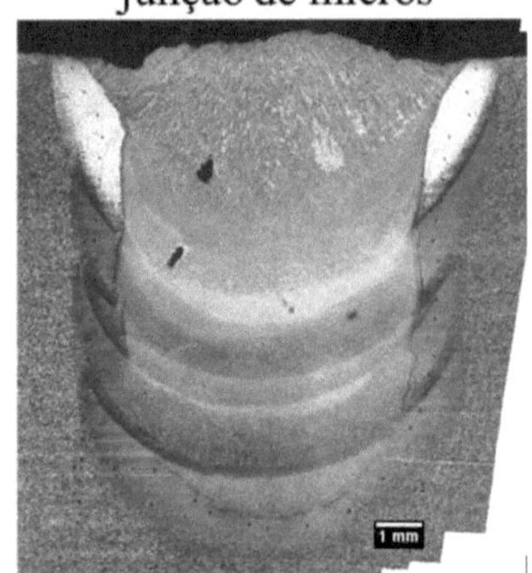

FIGURA 3.21   - Images of the same specimen. One taken by a conventional macrography camera, the other by joining several sequenced micrographs.

Figure 3.22 below is made up of more than 100 micrographs at 25x magnification.

40

This methodology was developed to allow the welding region to be analysed as a whole, making it possible to observe the grains and the different regions in the joint. Its use will be covered and discussed in the "Hardness evolution" chapter

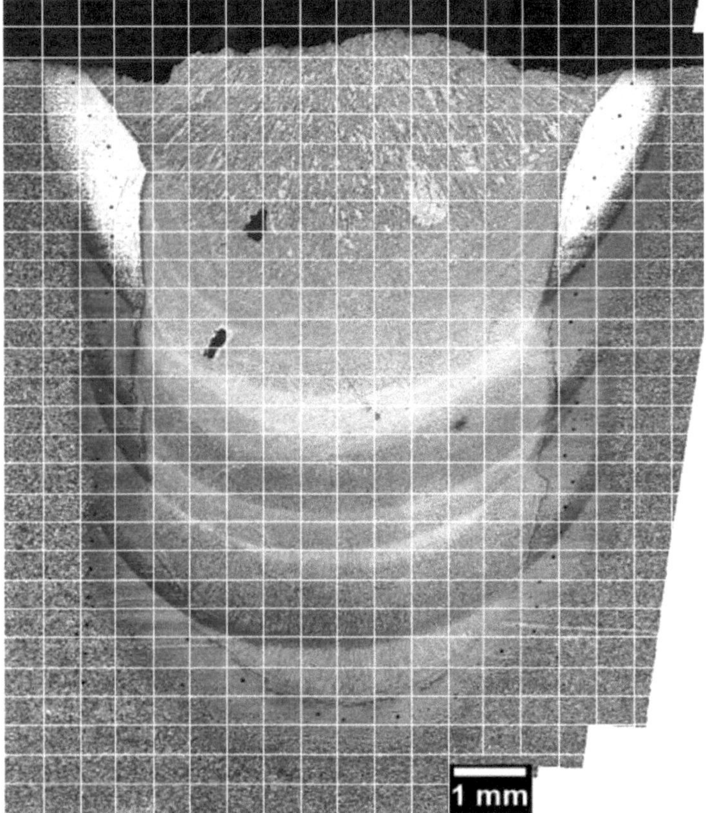

FIGURE 3.22 - Example of how a macrograph is assembled from micrographs.

3.8.   2Welding configurations   used

In order to study each technique's effect on the joint's quality, four different configurations of techniques will be studied on the test specimens:

-No   inversion and No grinding (S/I and S/E).

-With   inversion and Without grinding (C/I and S/E).

-Without   inversion and With grinding (S/I and C/E).

-With   inversion and With grinding (C/I and C/E).

To present the results for each configuration, the specimens (SPs) will be numbered from 1 to 8 following the criteria adopted in Table 3.2 below:

TABLE 3.2- Nomenclature for the specimens according to whether or not the techniques studied were used.

Configuration	CP number
No inversion / No grinding	1 e2
With inversion / Without grinding	3e4
Without inversion / With grinding	5e6
With inversion / With grinding	7e8

CHAPTER 4

RESULTS AND DISCUSSIONS

In this chapter, in addition to the results of the hardness tests and the study of the ZTAs for the 4 configurations used, the preliminary results of the methodology will be presented, such as the calculation of the welding energy used.

4.1 Welding parameters

To select the best welding parameters for the case of SAE 1045 base metal and E6013 electrode, six tests were carried out, varying the current and welding speed. All these tests were monitored by the acquisition system for subsequent calculation of the welding energy for each case.

The compared current and welding speed values were selected from internal research already carried out at the LRSS. Table 4.1 shows the parameters that were studied.

TABLE 4.1 - Current values and welding speeds studied before selecting the final parameterisation.

Welding angle (°)	Welding current (A)
65	160
65	200
60	160
60	200
55	160
55	200

With all the strands in the different patterns made, a visual analysis was made to check which parameterisation would suit the job. Figure 4.1 shows the strands all made at a depth of lOm.

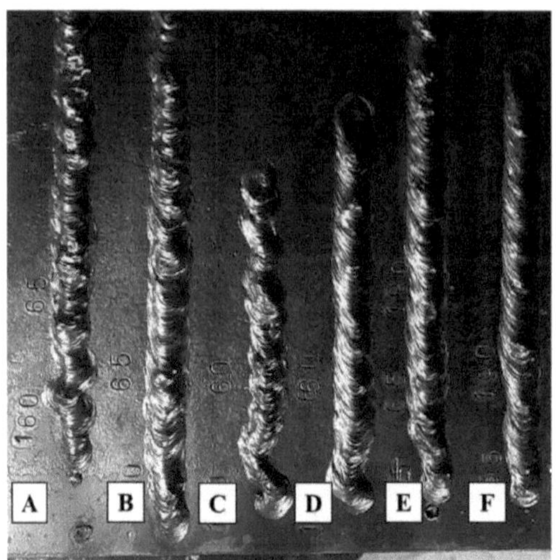

FIGURE 4.1 - Strands made to choose the parameters: (A)65°/160A; (B)657200A; (C)607160A; (D)60°/200A; (E)557160A; (F)557200A.

After visual analysis of the deposited strands, the parameterisation used was **60°** and **200A,** strand (D) in Figure 4.1.

4. 2Welding energy calculation for the process

In order to calculate the welding energy of the process, which is linked to the formation and characteristics of the HAZ, oscillograms were captured with the voltage and current values in real time during the tests at a rate of 5000 points per second. Figure 4.2 below shows the voltage and current oscillograms acquired at a welding angle of 60° and a current of 200A.

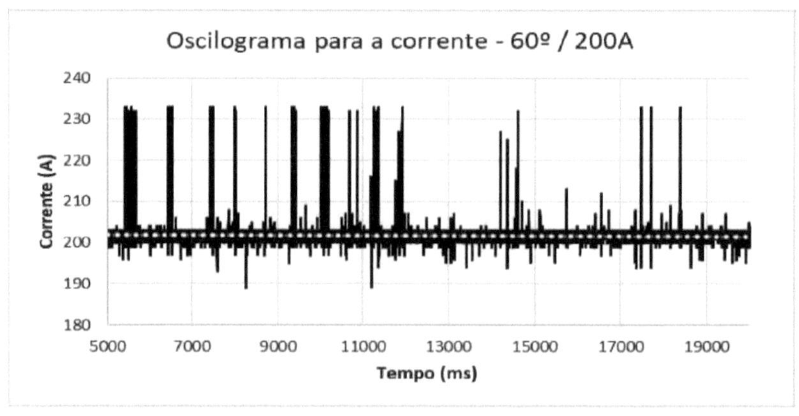

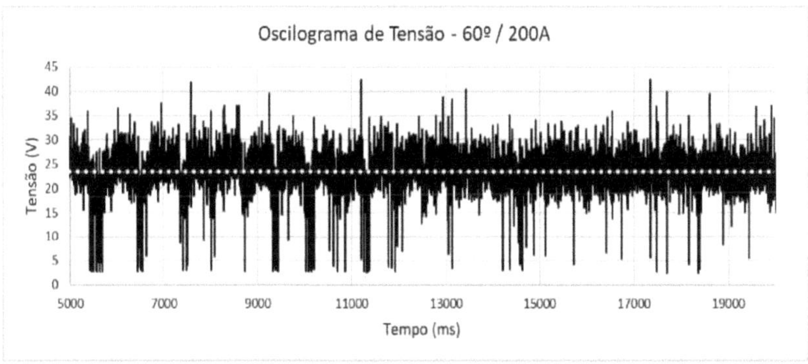

FIGURE 4.2 - Oscillograms of Welding Current and Welding Voltage. Parameters: 60° and 200A. Dotted line showing the trend of the curve.

Table 4.2 shows the values of the parameters used and the result of the welding energy as a function of the speed at which gravity welding was carried out. The average voltage and average current values were obtained using the "Signal" programme developed in the LRSS.

TABLE 4.2 - Welding parameters used to weld the chamfers.

Angle of Welding (°)	Average welding current (A)	Tension Average (V)	Speed Welding (mm/s)	Energy from Welding (kJ/mm)
60	202	23,7	3,88	1,23

43 Macrographs of the welded chamfers

The conventional macrographs of the welded chamfers are shown in Figures 43 4.4, 4.5 and 4.6, respecting the name adopted in Table 3.2.

No inversion / No grinding - CP 1 and CP 2

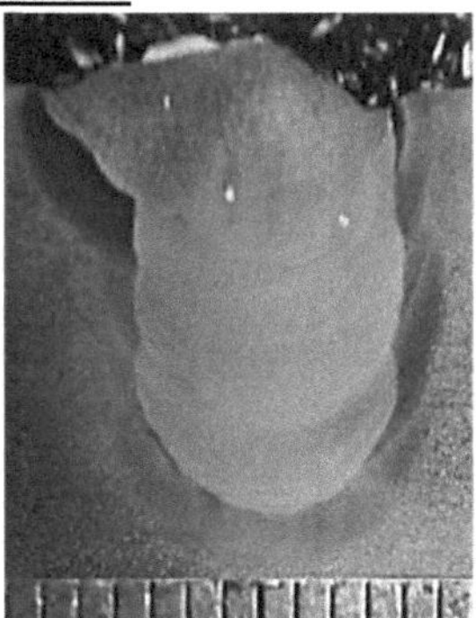

FIGURE 4.3 - Macrographs CP 1 and CP 2 - Configuration: No inversion / No grinding
With inversion / Without grinding - CP 3 and CP 4

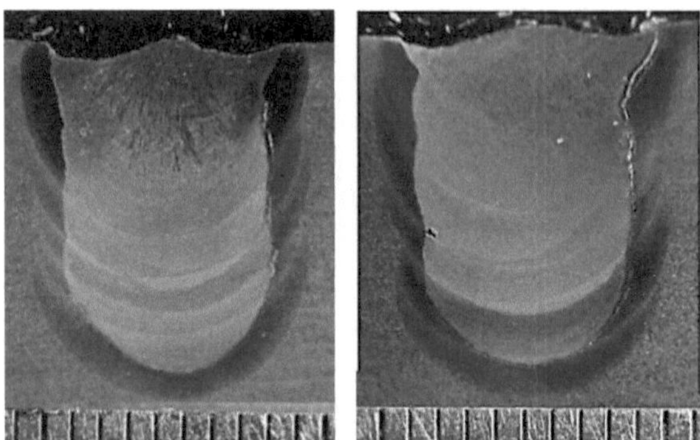

FIGURE 4.4 - Macrographs CP 3 and CP 4 - Configuration: With inversion / Without grinding

Without inversion / With grinding - CP 5 and CP 6

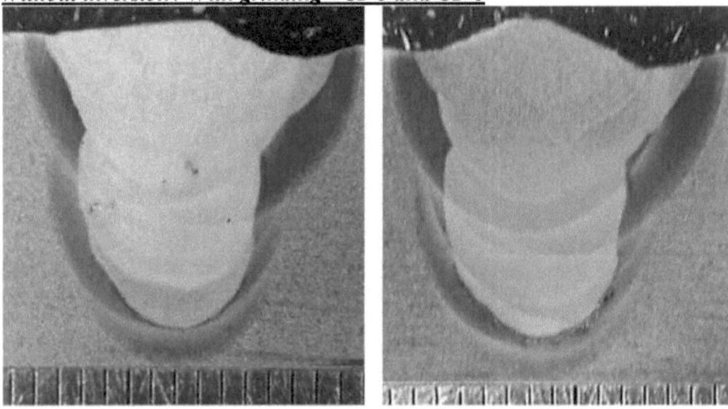

FIGURE 4.5 - Macrographs CP 5 and CP 6 - Configuration: Without inversion / With grinding.

With inversion / With grinding - CP 7 and CP 8

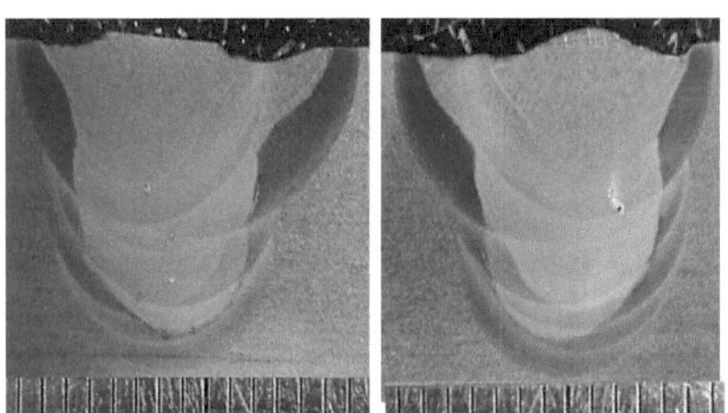

FIGURE 4.6 - Macrographs CP 7 and CP 8 - Configuration: With inversion / With grinding.

4.   4Hardness test results

The results for the hardness test of the filled chamfer are shown in the graphs below (figures 4.8 to 4.15), which show the hardness values for each side of the welding region. For comparison with the reference value stipulated by the AWSD3.6 standard, the reference line is in red, indicating 325 vickers.

Figure 4.7 shows the regions identified for the hardness test, this macrograph specifically of the CPI.

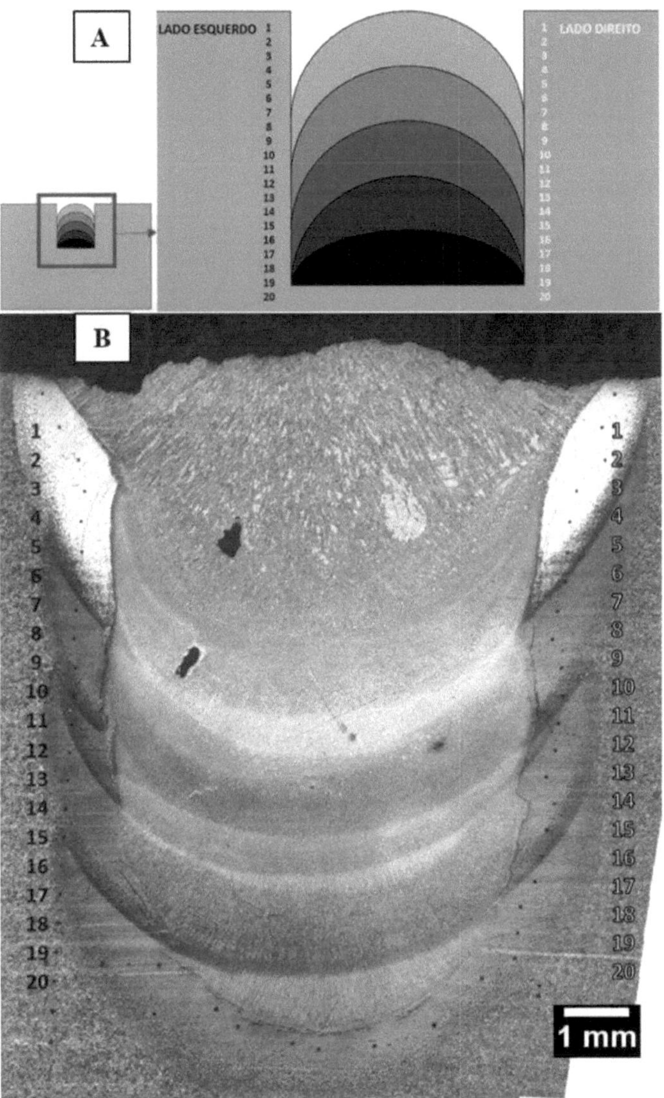

FIGURE 4.7- Positions of the indentations following the methodology. In (A) there is a schematic of the welded chamfer with the positions of the identifications; in (B) it is possible to visualise the identifications made on the welded specimen (CPI).

4.4.1 Without   Inversion / Without Grinding

The results of the hardness measurements for Specimens 1 and 2, which were not subjected to the direction reversal or grinding technique, are shown in Figures 4.8 and 4.9. To recap, the positions on each side are shown from 1 to 21, with their distribution starting from the surface of the chamfer (1) towards its interior (21), always following the HAZ. Figures 4.8 and 4.9 show the hardness profile of the HAZ for specimens 1 and 2, where no technique was applied.

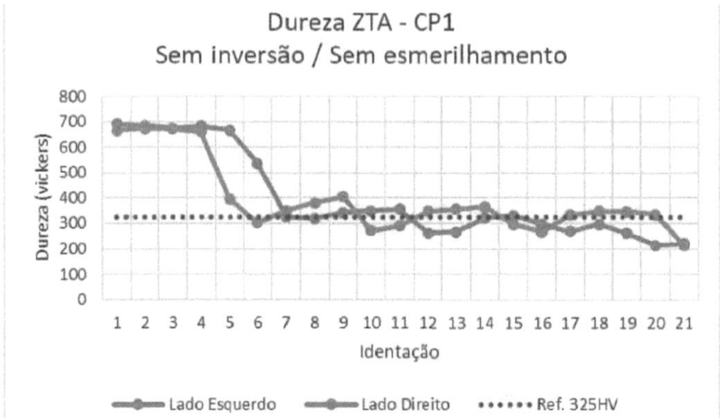

FIGURA 4.8   - . Hardness profile of the HAZ for the sides of the CPI specimen welded without reversing direction and without grinding. Label 1 - top of chamfer; Label 21 - MB.

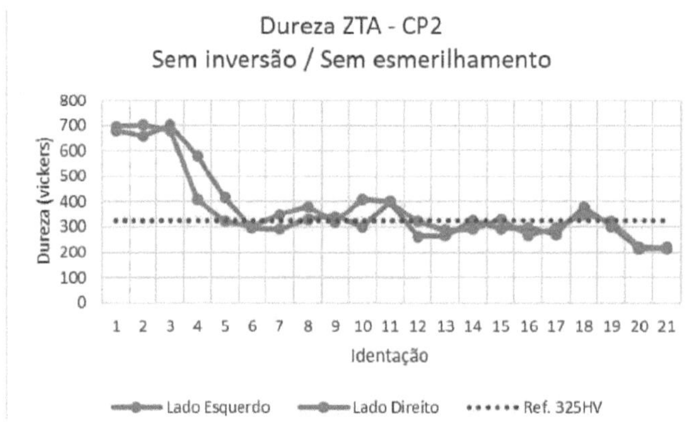

FIGURE 4.9 - Hardness profile of the HAZ for the sides of the CP2 specimen welded without reversing direction and without grinding. Label 1 - top of the chamfer; Label 21 - MB.

The diagrams presented show some hardness values above the established standard adopted as a reference in this work. The main regions showing these values are the surface regions, which is to be expected as the last pass does not undergo the possible tempering of a subsequent pass.

However, there are other values above the stipulated, and these are found in internal positions in the welded chamfer. They are listed in Table 4.3 below:

48

TABLE 4.3 - Internal regions with hardness above 325HV for CP 1 and CP 2.

	Excessive internal hardness (Above 325HV)	
	Left side	
CPI	9; 10; 11; 15	7; 8; 9; 12; 13; 14; 17; 18; 19; 20
CP 2	8; 9; 14; 18	7; 8; 10; 11; 15; 18; 19

A later subchapter will discuss these peaks, which are present in both samples of the same configuration.

4.4.2 With Inversion / Without Grinding

Figures 4.10 and 4.11 below show the diagrams with the hardness results for CPs 3 and 4. In this configuration, the joint passes are subjected to inversion in the welding direction, but are not subjected to the grinding method.

This configuration is important because different results to those presented in CP 1 and 2 will indicate how the inversion method influences the hardness values of the HAZ formed.

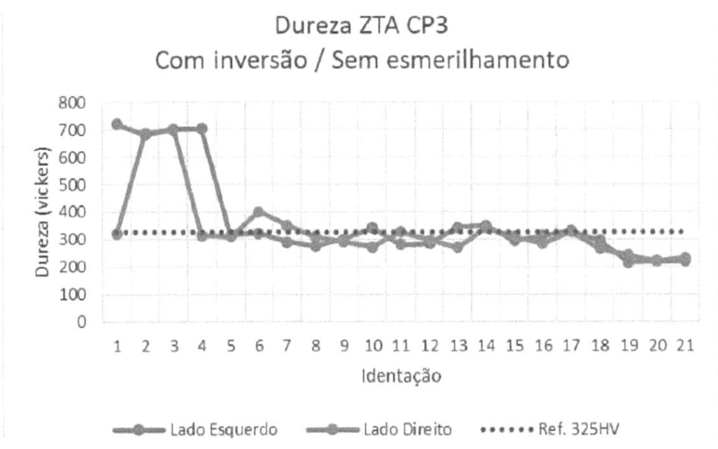

FIGURA 4.10 - Hardness profile of the HAZ for the sides of the CP3 specimen welded with inversion of direction and without grinding. Label 1 - top of chamfer; Label 21 - MB.

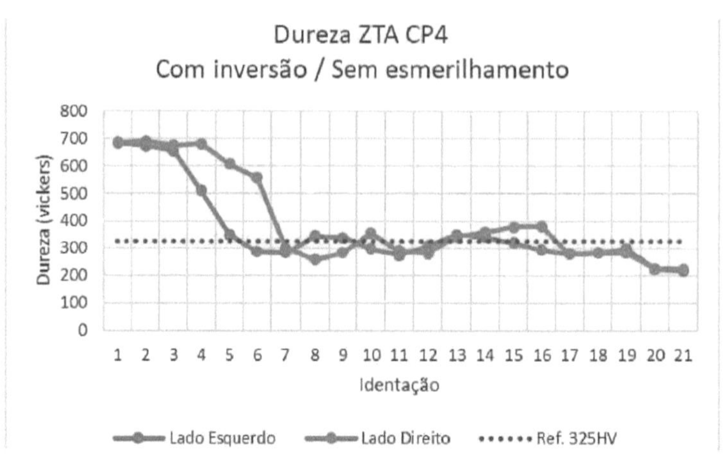

FIGURA 4.11   - Hardness profile of the HAZ for the sides of the CP4 specimen welded with inversion of direction and without grinding. Label 1 - top of chamfer; Label 21 - MB.

In the two CPs where the direction reversal technique is used and the grinding technique is not used, points of excessive hardness can be seen on the surface and in the inner regions of the chamfer. Table 4.4 below shows the indentations where the internal regions of the chamfer showed hardness above 325 HV.

TABLE 4.4 - Internal regions with hardness above 325HV for CP 3 and CP 4.

	Internal regions of excessive hardness (Above 325HV)	
	Left side	
CP 3	10; 13; 14; 17	6; 7; 11; 14; 15; 17
CP 4	8; 9; 13; 14;	10; 13; 14;15;16

443 Without Inversion / With Grinding

CPs 5 and 6 went through the grinding process, but the welding direction was maintained in all passes, so comparing these joints with CPs 1 and 2 will show how the grinding process influences the properties of the welded joint.

Figures 4.12 and 4.13 below show the hardness diagrams for bevels CP5 and CP6, where 7 passes were made. The number of weld beads when the grinding process is used is greater due to the removal of material with each pass.

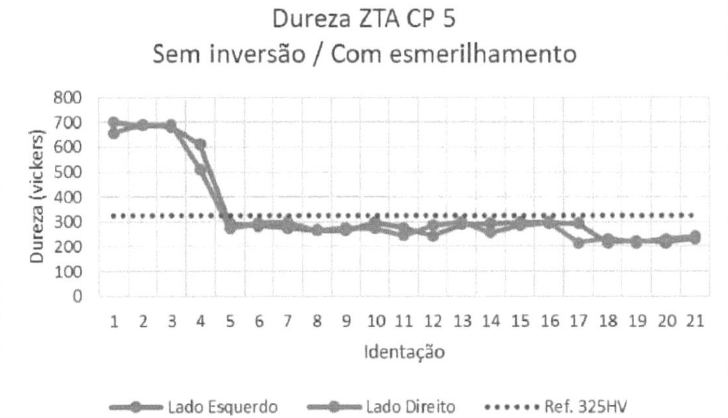

FIGURA 4.12   - Hardness profile of the HAZ for the sides of the CP5 specimen welded without reversal of direction and with grinding. Label 1 - top of chamfer; Label 21 - MB.

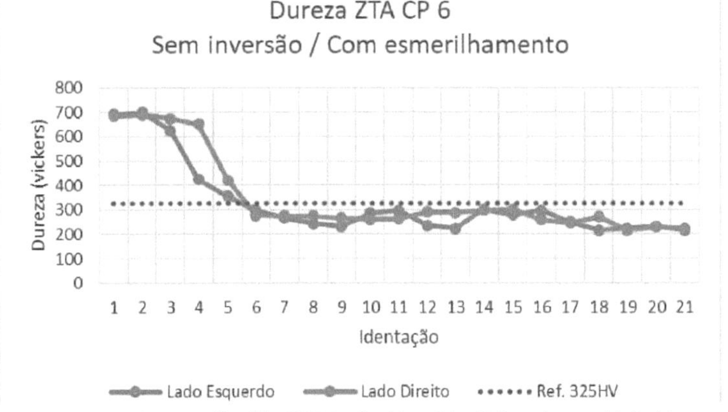

FIGURE 4.13 - Hardness profile of the HAZ for the sides of the CP6 specimen welded without reversal of direction and with spalling. Label 1 - top of the chamfer; Label 21 - MB.

The absence of regions of excessive hardness inside the chamfer can be seen for these specimens, as well as a greater homogeneity of values for the HAZ of the internal regions of the welded joint, and these values remain below the value stipulated as the maximum by the standard, which is 325HV.

Although there are no regions of excessive hardness inside the chamfer, the surface region of the HAZ shows values of around 600HV.

Table 4.5 below follows the same methodology as the tables presented for CPs 1, 2, 3 and 4, but as no regions of excessive hardness were recorded, the table does not show any indentation.

TABLE 4.5 - For CPs 5 and 6 there was no internal region with a hardness above 325HV.

Internal regions of excessive hardness (Above 325HV)	
Left side	

| CPS | N/A | N/A |
| CP 6 | N/A | N/A |

4.4.4 With Inversion / With Grinding

The fourth configuration used in this work was to work with the inversion technique and grinding simultaneously, i.e. in CPs 7 and 8 the chamfers were subjected to the two techniques studied.

Thus:

- Compared to CPs 3 and 4, the role of grinding on the properties of the joint will be evaluated.

- If compared with CPs 5 and 6, the role of the inversion of direction on the properties of the joints will be evaluated.

Figures 4.14 and 4.15 below show the diagrams with the hardness results for CPs 7 and 8, which were subjected to the direction reversal and grinding techniques.

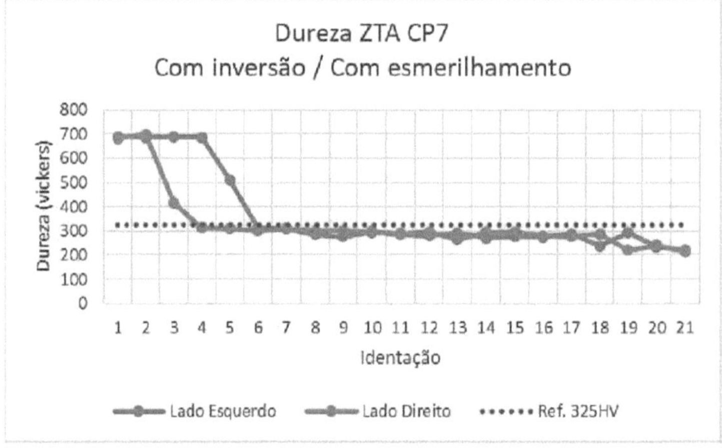

FIGURA 4.14   - Hardness profile of the HAZ for the sides of the CP 7 specimen welded with inversion of direction and spalling. Label 1 - top of chamfer; Label 21 - MB.

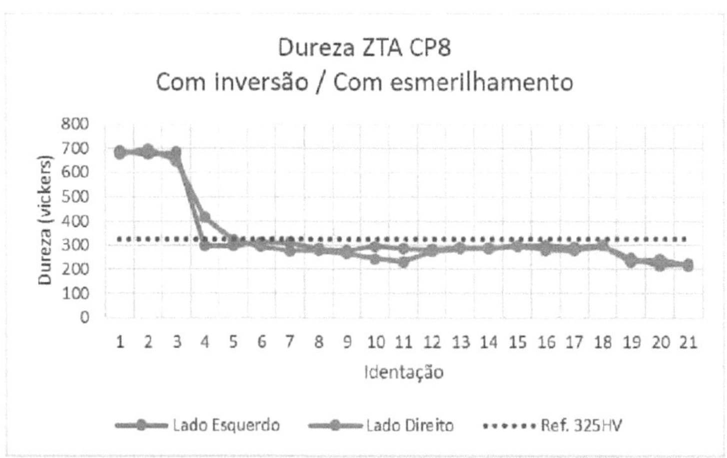

FIGURE 4.15 - Hardness profile of the HAZ for the sides of the CP 8 specimen welded with inversion of direction and with reaming. Label 1 - top of the chamfer; Label 21 - MB.

In the diagrams above *it* can be seen that CPs 7 and 8 also show excessive hardness in the surface regions of the HAZ, but like CPs 5 and 6 they do not show hardness above 325HV in the internal regions of the welded joints.

Table 4.6 below shows the method of presenting the internal regions with excessive hardness for CPs 7 and 8

TABLE 4.6 - For CPs 7 and 8 there was no internal region with a hardness above 325HV.

	Internal regions of excessive hardness (Above 325HV)	
	Left side	
CP7	N/A	N/A
CP8	N/A	N/A

445 General Result

For a better understanding and reading of the diagrams, we can separate the results into 2 types, those with high hardnesses inside the HAZ and those without these values above the norm, as shown in Table 4.7:

TABLE 4.7 - Results of the presence or absence of excessively hard regions inside the chamfers according to the techniques used.

Configuration			Resultados	
CPS	Inverting	Grinding	Excessive hardness in internal regions	Excessive surface hardness
1 e2				
3e4	YES			
5e6		YES		
7e8	YES	YES		

It can be seen that the specimens that underwent the grinding process did not show excessive hardness in the internal regions of the HAZ, but all the samples showed excessive hardness in the regions of the HAZ close to the surface. This means that the grinding method prevented hardness values above those considered ideal by the standard from being detected in the internal regions of the samples in which this procedure was adopted, but, as expected, it did not influence the more superficial regions of the HAZ, which continued to show high hardness.

43 "Evolution of hardness"

In order to better understand how the hardness behaves in the chamfers in which there was no grinding, and to investigate the reason for the high hardness peaks in the internal regions of the HAZ, the hardnesses were analysed in each welding pass. The "slices" taken from each pass were subjected to the methodology presented in which macros are made from micros. The specimen selected was specimen 1 (CPI) and each slice is shown below, where you can see the grains and the different regions of the HAZ.Figures 4.16, 4.17, 4.18, 4.19 and 4.20 show the macrographs made from micrographs. Hardness tests were carried out on the slices following the same methodology presented above. The hardness results are presented below.

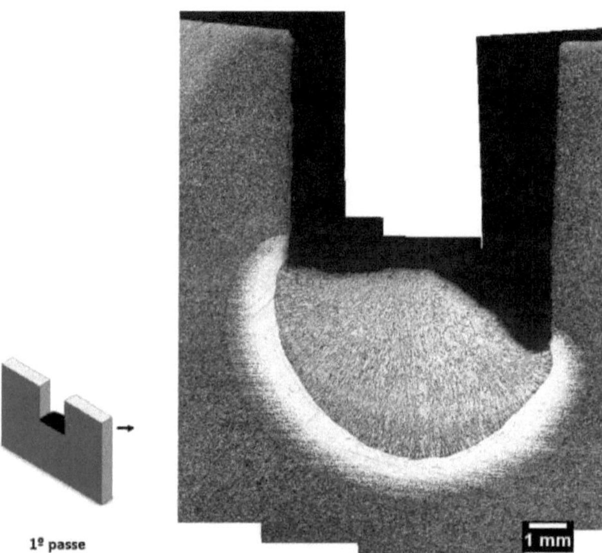

FIGURA 4.16-   - Schematic of the deposition of the Iº pass for CP 1 next to the macrograph made from micrographs.

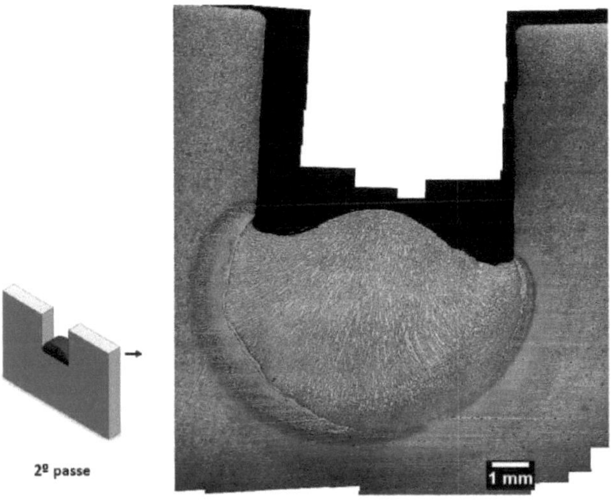

FIGURA 4.17-   Diagram of the deposition of the 2° pass for CP 1 next to the macrograph made from micrographs.

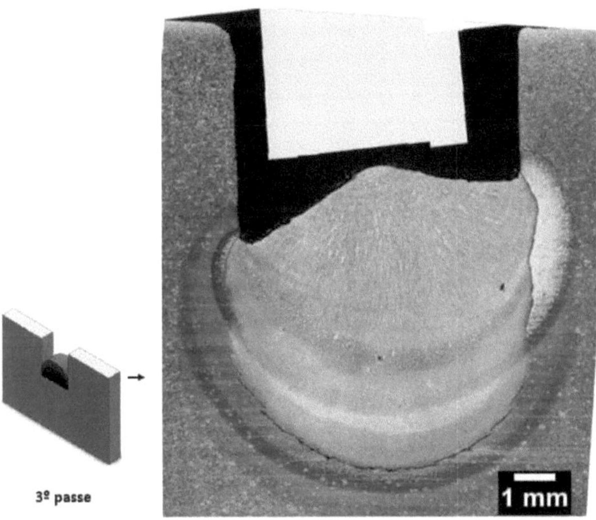

FIGURA 4.18-   Schematic of the deposition of the 3° pass for CP 1 next to the macrograph made from micrographs.

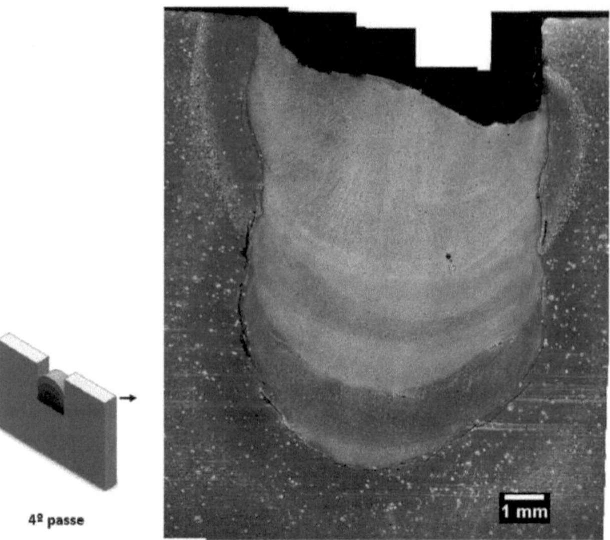

FIGURA 4.19-   Diagram of the deposition of the 4° pass for CP 1 next to the macrograph made from micrographs.

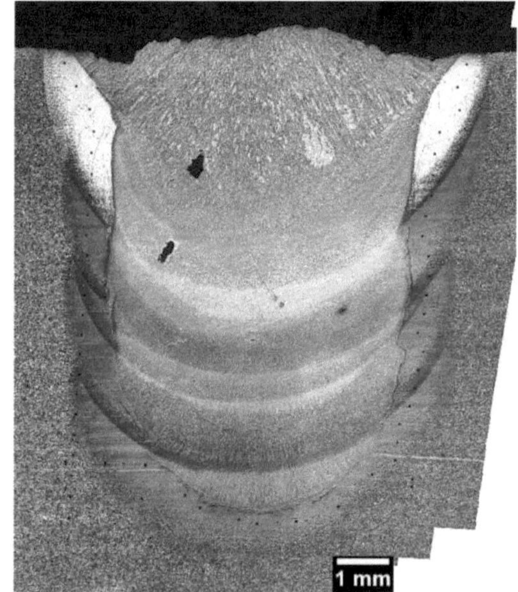

FIGURE 4.20- Schematic of the deposition of the 5° pass for CP 1 next to the macrograph made from micrographs.

Following the same methodology as for the hardness test, the graphs below show the **evolution of the hard zone, which is** always present in the uppermost bead of the chamfer. This hard zone is characterised by values in the 500 to 600 Vickers range.

The diagrams below are shown next to the slices where the identifications were made, so it is possible to identify the regions with high hardness next to the image.

Figure 4.21 shows the hardness values for CP 1, when only 1 welding pass had been made. The surface identifications (1 to 12) therefore show hardness values corresponding to the hardness of the base metal without the influence of the energy imposed by the welding process.

It is also possible to see two regions of high hardness in the innermost indentations of the chamfer (indentations 15 to 20). These regions of high hardness are the result of the rapid cooling of the HAZ at the time of welding.

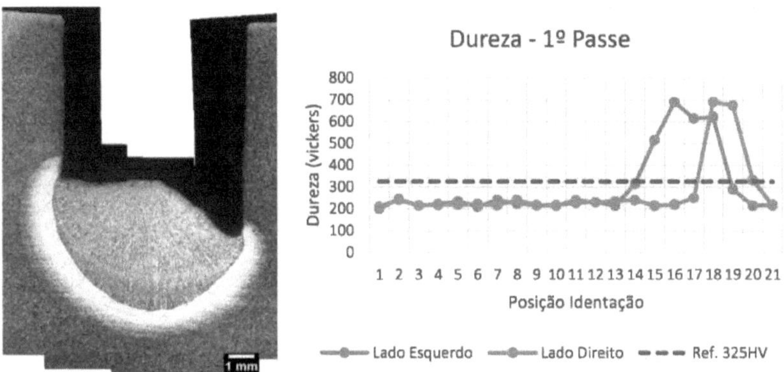

FIGURE 4.21 - Macro of the I° pass of CP 1 next to its hardness diagram.

Figure 4.21 also shows a difference in the position of the hardness peaks. This difference is a consequence of non-symmetry in the deposition, generating a difference in heat flow to the base metal and leaving the HAZ with larger dimensions on one side of the chamfer. In the case above, the HAZ on the left (the reader's) side is larger and reaches higher regions (identifications 14 to 18) while the HAZ on the right is smaller (18, 19 and 20).

Figure 4.22 below shows an HAZ with internal regions of high hardness on the left-hand side, but no such values on the right-hand side. The image shows that the HAZ of the 2nd° pass was concentrated on the left, explaining the absence of a very hard zone on the right.

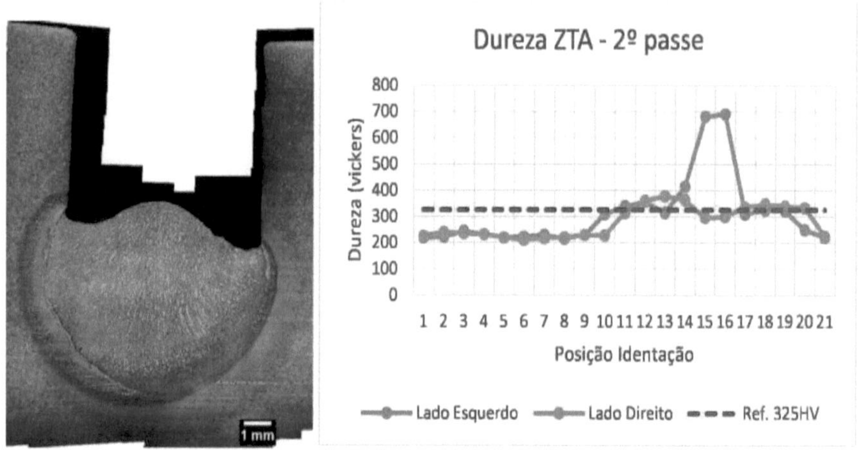

FIGURA 4.22-   Macro of the 2nd° pass of CP 1 next to its hardness diagram.

Figure 4.23 shows the same phenomenon explained in Figure 4.13, with a ZTA concentrated on one side.

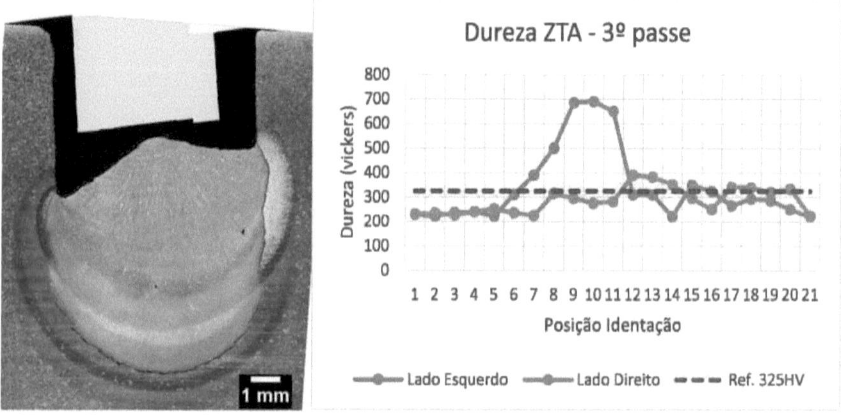

FIGURA 4.23-   Macro of the 3° pass of CP 1 next to its hardness diagram.

Figure 4.24 shows two hardness peaks again, close to the surface. The image shows that the HAZ on the right has already reached the surface, while the HAZ on the left is still in the inner regions of the chamfer, explaining the low hardness of identifications 1 and 2 for this side, with these hardness values corresponding to the base metal.

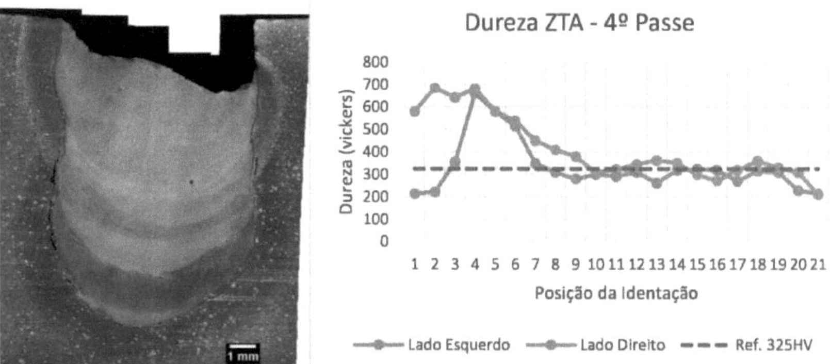

FIGURA 4.24-   Macro of the 4th° pass of CP 1 next to its hardness diagram.

Figure 4.25 shows the same diagram as in Figure 4.8. The image on the right improves understanding and visualisation of the diagram. The hardness peaks on the surface are due to the last HAZ being positioned there. The internal regions show hardness peaks, which are not acceptable by the norm.

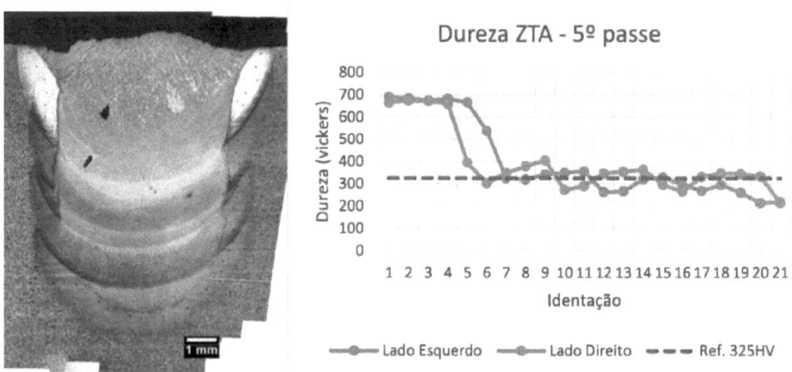

FIGURE 4.25- Macro of the 5° pass of CP 1 next to its hardness diagram.

In order to understand the evolution of hardness, the graphs below show separate values for each side and each pass. We have taken into account the position of the identifications that show values above those recommended by the standard, and we have analysed the history of these points considered to be "faulty", as explained in the graphs below.

Figure 4.26 shows the hardness value diagrams for the left-hand side only. The graphs are placed so that the positions of the identifications (1 to 21) are "stacked".

This analysis will explain the presence of **hardness peaks** in the inner regions of the chamfer. This explanation will be given using the hardness diagram of previous passes, comparing the same points but for different quantities of weld beads deposited.

The aim is to explain how the hardness in the different regions of the chamfer was influenced by the heat imposed by subsequent passes.

59

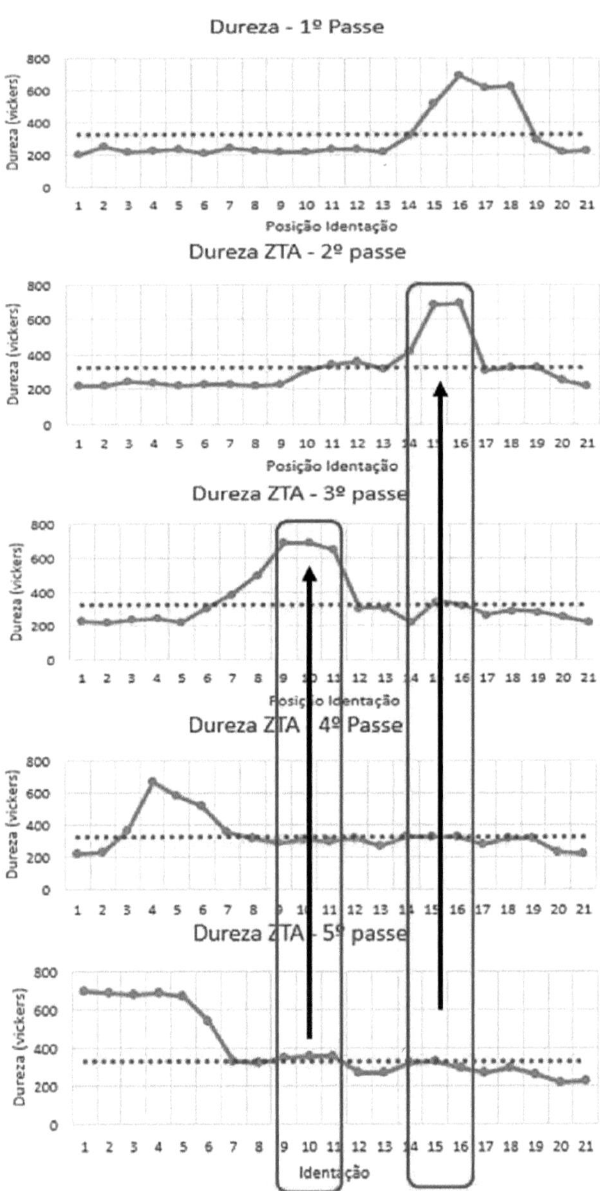

FIGURA 4.26-   The black arrows point to the areas that, in previous passes, had high hardness and that, even after subsequent passes, remained with high hardness until the last pass.

Considering the diagrams superimposed with the hardness values for the HAZ on the left side of CP 1 (without grinding), it can be seen that:

In the 5º pass, disregarding the surface ZTA region, the regions that showed excessive hardness are regions where, in previous passes, the surface ZTA was of high hardness, i.e. regions of 500 to 600 HV.

60

The black arrows indicate the "origin" of the excessive hardness, they show that in some pass the hardness was extremely high, and that not all of this region underwent adequate tempering in subsequent passes, remaining with excessive hardness values.

It can be concluded that the internal regions in the final joint that show unacceptable values are regions that **have not undergone adequate tempering,** keeping some regions excessively hard. The same is observed for the right side of the chamfer.

The right-hand side also shows internal regions of high hardness. These points coincide with regions that were once a "surface HAZ".

The fact that the hardness remains higher than desirable shows that the heat imposed by the subsequent passes is not sufficient to temper the entire length of the HAZ, leaving hard regions.

Figure 4.27 shows **the hardness evolution** for the right side of the CPI chamfer.

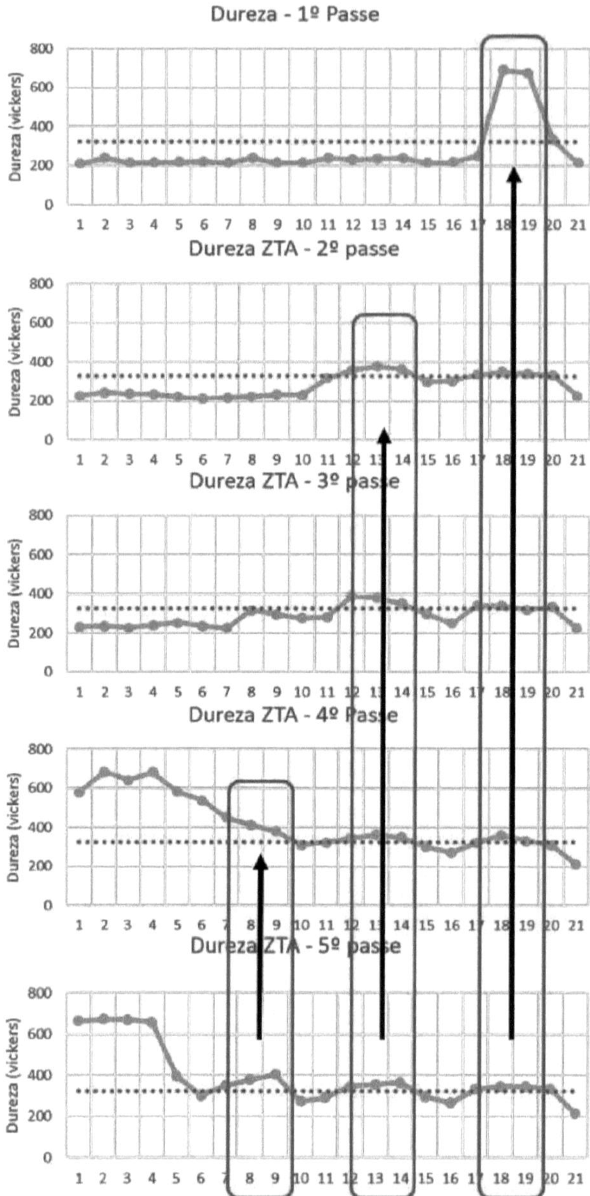

FIGURA 4.27-   The black arrows point to the areas that, in previous passes, had high hardness and that, even after subsequent passes, remained with high hardness until the last pass.

As the specimens that were subjected to the grinding process did not show internal regions with high hardness, it can also be said that they underwent an appropriate tempering process, as the use of heat is more efficient, since the ZTAs of the passes made are closer together due to the removal of part of the reinforcement.

62

4.6 Effect of tempering

With the methodology of obtaining macros from micros, it is possible to analyse a specific region with its indentations, and thus better observe the effect of tempering the HAZ in multipass welding. The figures below show the difference in the size of the indentations between the surface region, which has around 600HV, and the tempered regions, with values below 325HV. Figure 4.28 shows macrographs highlighting a specific region of the HAZ, the right side of the CPI, where the region analysed is identified in red.

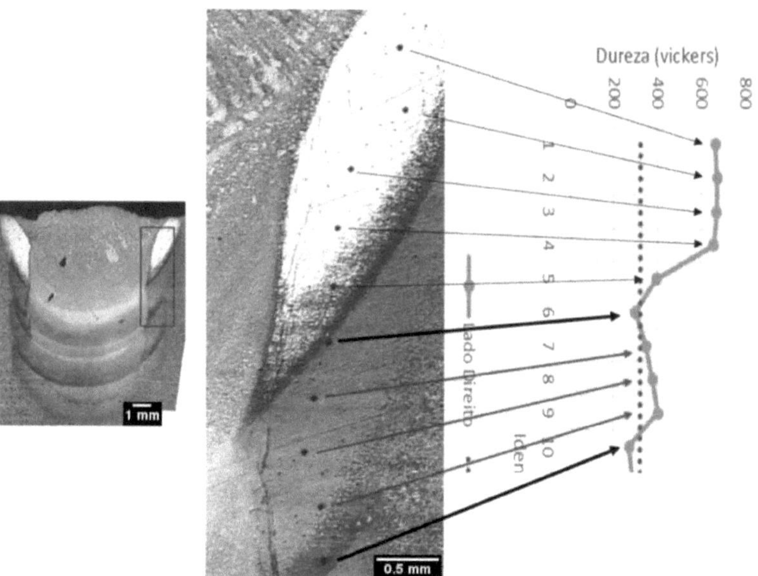

FIGURE 4.28- Macrograph of CP 1 on the left with the red region highlighted. The surface and internal region of the HAZ are highlighted, showing the difference in hardness between the identifications.

Firstly, the smallest size is noticeable for identifications 1, 2, 3 and 4, which are located in the surface HAZ region.

The image also shows the high hardness due to the lack of tempering in identifications 7, 8 and 9.

You can also see the regions where tempering was effective, considerably reducing the hardness values, which are in the regions of indentation 5, 6, and 10, for this range presented.

Figure 4.29 below shows CP 4, which has not undergone the grinding process either, although it has been reversed. In the region highlighted in red, it is possible to see the indentations and the lack of proper tempering is noticeable in indentations 10, 13, 14, 15 and 16, marked by red arrows.

The regions indicated by the black arrows and labelled 11, 12 and 17 have hardness values below 300HV and the image shows that they have been effectively tempered, which reflects on the property of these regions.

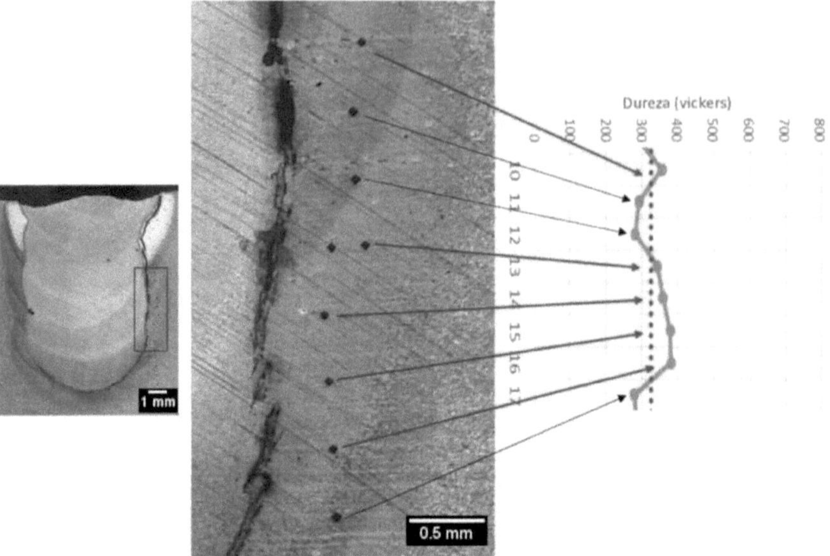

FIGURE 4.29- Macrograph of CP 4 on the left with the red region highlighted. The difference in hardness between the identifications shows the dependence of the degree of tempering on this property.

4.6.1 Effect of   tempering with the grinding technique.

Applying the same methodology to specimens that have undergone grinding, it can be seen that the absence of high hardness peaks is due to the effect of more effective tempering. Tempering is more intense in these specimens due to the presence of a greater number of ZTAs, i.e. there are regions that have reached higher temperatures closer together.

Figure 4.30 shows the macrograph of CP 5, which only underwent the grinding process and shows the HAZ of the last pass, basically martensitic and with high hardness values (700 to 600HV) indicated by the red arrow.

Just below this region, a darker colour is noticeable along the entire internal length of the HAZ. This is because these regions underwent significant heating in the subsequent pass, thus undergoing a high degree of tempering, reflected in the darker colour due to the release of carbon that was in solid solution. This high degree of tempering is reflected in the decrease in hardness values, with values below 325 vickers being measured.

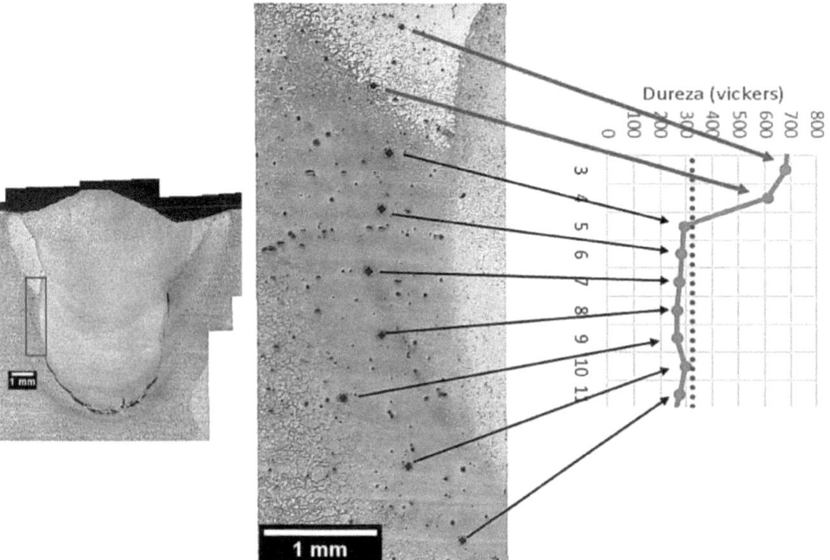

Figure 4.30 Macrograph of CP 5 on the left with the red region highlighted. High hardness values for the surface HAZ. Internal regions with darker colouring and lower hardness values, the effect of tempering.

CHAPTER 5

CONCLUSIONS

Based on the results and discussions presented in this work, the main conclusions are:

1. The grinding method proved to be effective in preventing excess hardness in internal regions in a multi-pass weld, but it increases welding time due to the preparation for each pass and also increases the number of passes required.
2. The direction reversal method did not have a positive influence on the results obtained for tempering the HAZ and therefore showed regions of hardness above the values established by the standard. The same is true of multipass welds which do not use the direction reversal method or grinding.
3. The opening chamfer as used is a good method for studying the HAZ in multipass underwater welds.
4. The application of the controlled grinding technique can significantly improve some characteristics of joints produced by wet welding, helping to achieve the results required to qualify for class A wet welding procedures, according to AWS D3.6M.

SUGGESTIONS FOR FUTURE WORK

Future work is proposed:

1. Carry out impact and/or fatigue tests to measure the toughness of the HAZs produced using each method.

2. Create a methodology for analysing the properties of the molten zone for the grinding and inversion methods.

3. Use a similar chamfer but with the presence of a backing, in order to eliminate the formation/propagation of a cold crack that appears in the lower part of the first pass.

4. Apply the temper bead method to similar joints and evaluate the reduction in hardness in the HAZ near the surface.

5. To study the effect on the mechanical properties of multi-pass underwater welding using other post-weld heat treatment methods, such as the use of electromagnetic induction.

6. Applying the techniques studied to "V" bevels within the established standard for producing specimens, thus validating the methods as a tool for improving the mechanical properties of wet underwater welds.

BIBLIOGRAPHICAL REFERENCES

AGUIAR, W.M. "Welding ABNT 4140 steel without subsequent heat treatment". Master's thesis - Materials Science and Engineering Programme, Federal University of Ceará. Fortaleza, 2001.

AMERICAN WELDING SOCIETY. Underwater Welding Code. ANSI/AWS D3.6M:2010.

ANDRADE, L. G. D. "Efeito do Teor de Carbono do Metal Adição e do Metal de Base Sobre a Porosidade do Metal de Solda Subaquática", Belo Horizonte, Universidade Federal de Minas Gerais, lOlf, 2010;

ARRAES JÚNIOR, R. M. "Evaluation of the Tenacity of the ZAC in the Welding of ABNT 4340 Steel without Subsequent Heat Treatment". Fortaleza, 2001. Master's thesis - Materials Science and Engineering Programme, Federal University of Ceará.

ASM Metals Handbook - Volume 06 -Welding, Brazing and Soldering, TA459.M43 1990 620.1'6 90-115; ISBN 0-87170-377-7(V.l), SAN 204-7586 ISBN 0-87170-382-3, Printed in THE UNITED STATES OF AMERICA, 1993.

ASME Boiler and Pressure Vessel Code, Section XI, Article IWB-4000, Repair Procedures. New York, 1995.

BRAZILIAN ASSOCIATION OF TECHNICAL STANDARDS (ABNT). Carbon and Alloy Steels for Mechanical Construction - Designation and Chemical Composition, NBR NM 87/2000. Rio de Janeiro, 2000.

BRAZILIAN ASSOCIATION OF TECHNICAL STANDARDS (ABNT). Steel Classification Criteria, NBR NM 172/2000. Rio de Janeiro, 2000.

BRAZILIAN ASSOCIATION OF TECHNICAL STANDARDS (ABNT). Metallic Materials - Vickers Hardness. Part 1 - Vickers hardness measurement. NBR NM 188-1. Rio de Janeiro. May. 1999.

AWS D3.6M: 1999: Specification for underwater welding, American Welding Society, Miami, USA 1999.

BAILEY, N. "Weldability of ferritic steels". England: Abington Publishing, 1994, 275p.

BRACARENSE. A. Q., PESSOA. E. C. P., SANTOS. V. R., MONTEIRO. M. J., RIZZO. F. C.. PACIORNIK. S.. REPPOLD. R., DOMINGUES. J. R., VIEIRA. L. A. 2008. "Comparative study of commercial electrodes for underwater wet welding". HW International Congress - 2nd Latin American Welding Congress XXXIV CONSOLDA - Congresso Nacional de Soldagem. São Paulo. SP. 2008.

CHRISTENSEN N. The Metallurgy of MMA Hyperbaric Welding, SINTEF report N°. STF34 F83032, Trondheim, 1983.

COE, F.R. "Welding steels without hydrogen cracking". The Welding Institute, Cambridge, 1973. 68 p.ll. NACE. Basic coirosion course. Anton de S. Brasunas. Houston- Texas, llth ed., June 1990.

DAVID, S. A., VITEK, J. M. "Correlation between Solidification Parameters and weld microstructures". International Materials Reviews. Vol. 34, No. 5, pp. 213-245, 1989.

FILHO, J.C.P; MELLO, R.T.; MEDEIROS, R.C.; PARANHOS, R. Recent History of Wet Underwater Welding, 2004;

FUKUDA T., SUMIYA R., KONO W., SUEZONO N., TAMURA M., and CHIDA, I. "Temper- Bead Weld by Underwater Laser Beam Welding", 17th International Conference on Nuclear Engineering, Brussels, Belgium, July 12-16, 2009.

GRANJON, H. 1972 "La Fissuration à Froid en Soudage D'acieres. Soud". Tec. Conn, 26 (3/4), Mar/1972. pl55-164.

GRAVILLE, B. A. "Cold Cracking in Welds in HSLA Steels". Welding of HSLA (microalloyed) Structural Steels, Proc. Int. Conf., American Society for Metals, 1976

GUERRERO, F. P. "Effect of nickel additions on rutile electrodes for underwater welding". Master's thesis - Metallurgical and Materials Engineering Department, ColoradoSchool of Mines, Golden, Colorado, USA, 84f, 2002.

H.T. Zhang, X.Y. Dai, J.C. Feng, L.L. Hu. "Preliminary investigation on real-time induction heating-assisted underwater wet welding". Weld. Journal, 94 (2015), pp. 8-15

HASUI, A., AND SUGA, Y. 1980. "Oncooling ofunderwater welds". Transactions of the Japan Welding Society 11: 21-28.

HIGUCHI, M.; SAKAMOTO, H.; TANIOKA, S. "A study on Weld Repair Through Half Bead Method". IHI Engineering Review. V.13, April 1980.

IBARRA, S., GRUBBS, C. E., LIU, S. "State-of-the-Art and Practice of Underwater Wet Welding of Steel", Proceedings: Intemationaal Workshop on Underwater Welding of Marine Structures. New Orleans, Lousiana, 1994. pp 49-67.

J. LABANOWSKI, D. FYDRYCH, G. ROGALSKI. "Underwater Welding - A Review". Poland, Gdansk, 2008

KOU, S. "Welding Metallurgy", A Wiley-Interscience Publication, 2th, USA, 2003;

KOU, Sindo. "Welding Metallurgy". led. New York: John Wiley & Sons, 1987. 41 Ip

KÚCHLER, M. M. - "Application of the Double Layer Technique in the Welding of Pipelines in Operation". Porto Alegre: UFRGS, 2009. 95 pg. Dissertation (master's degree). Postgraduate Programme in Mining, Metallurgical and Materials Engineering - PPGEM - Porto Alegre, RS -2009

LIU S., POPE, A.M., AND DAEMEN, R., "Welding Consumables and Weldability", International Workshop on Underwater Welding of Marine Structures, Lousiana, USA, 1994. pp.321-350.

MARINHO, R. R.; PAES, M. T. P.; PESSOA, E. C. P.; BRACARENSE, A. Q.; SANTOS, V. R.; ASSUNÇÃO, C. R.; MONTEIRO, M. J.; DOMINGUES, J. R. "Perspectives and Challenges for the Application of Wet Underwater Welding at PETROBRAS". Rio Welding. Rio de Janeiro, 2014

MAZZAFERRO, J. A. E. "Study of the Stability of the Electric Arc in Underwater Welding with Coated Electrodes". Doctoral thesis. Federal University of Rio Grande do Sul. Brazil, 1998.

METALS HANDBOOK. "Heat Treating of Steels. lOed. Ohio: American Society for Metals, v.4,1991

METALS HANDBOOK. "Selection of Carbon and Low Alloy Steel. lOed. Ohio: American Society for

Metals, v.6, 1992.

MODENESI, P. J., MARQUES, P. V., SANTOS, D. B. Curso de Metalurgia da Soldagem. Belo Horizonte: UFMG, 1992. 297p.

NACE. "Basic corrosion course". Anton de S. Brasunas. Houston-Texas, llth ed., June 1990.

NINO, C. E. B. "Specification of welding repair procedures without subsequent heat treatment - Tempering effect produced by thermal cycles". PhD thesis, Postgraduate Programme in Mechanical Engineering, Federal University of Santa Catarina.

OLSEN, K. et al. "Weld bead Tempering of the Heat-Affected Zone". Scandinavian. Journal of Metallutgy, nov/1982. p. 163-168.

PESSOA E. C. P., BRACARENSE A. Q., DOS SANTOS V. R., MONTEIRO M. D. J., VIEIRA L. A., MARINHO R. R.: "Challenges to develop an underwater wet welding electrode for "class A welds" classification, as required in the AWS D3.6 code". ASM Proceedings of the International Conference: Trends in Welding Research 2013, pp. 259

PESSOA, E. C. P., BRACARENSE, A. Q., LIU, S., GUERRERO, F. P., "Study of Re-Melt Temper Bead and Polarity Effects on Porosity in the Under Freshwater Wet Welds". 23rd International Conference on Offshore Mechanics and Arctic Engineering, 2004, Vancouver. Proceedings of OMAE04

PESSOA, E. C. P., BRACARENSE, A. Q., SANTOS. V. R., MONTEIRO. M. J., RIZZO. F. C., MARINHO, R. R., VIEIRA. L. A., SILVA, D. B. 2013. "Wet Welding Field Trials In Shallow Waters For Structural Repairs In Floating Oil Production Units". OMAE 2013 June 9- 14, 2013, Nantes, France.

PESSOA, E.C.P. "Estudo da Variação da Porosidade ao Longo do Cordão em Soldas Subaquáticas Molhados", Federal University of Minas Gerais, Belo Horizonte, Brazil, Thesis, 157f, 2007.

POPE, A.M., LIU, S. "Hydrogen Content of Underwater Wet Welds Deposited by Rutile and Oxidizing Electrodes", OMAE, Volume III, Mechanical Engineering, 1996;

POPE, A.M.; MEDEIROS, R.C.; LIU, S. "Solification of Underwater Wet Welds". Proceedings of the 14th International Conference on Offshore Mechanics and Artic Engineering, Material Engineering, Vol. III, pp. 54-63, April, USA, 1995.

SAVAGE W. F., NIPPES, E. F., AND SZEKERES, E. S. Weld. Joints, 55: 276s, 1976;

SHEWMON, P. "Transformations in metals". McGraw-Hill, USA, 1969, 394 p.

SILVA, W.C.D; PESSOA, E.C.P.; BRACARENSE, A.Q.; RIBEIRO, L.F.; ÁVILA, T. "Diffusible Hydrogen on Underwater Wet Welds Produced With Tubular Shielded Electrodes Using Internal Gas Protection", Natal, RN, Brazil, 21st Brazilian Congress of Mechanical Engineering, 2011;

SPERKO, W. J. "Exploring temper bead". Welding Journal - July 2005 - pag. 37 to 40 Szelagowski, P., and Ibarra, S. 1992. In-situ post-weld heat treatment of wetwelds. Offshore Technology Conference, Houston, Texas.

TEIXEIRA, J.C.G.; POPE, AM. "Double layer deposition technique for repairs and modifications without post-weld heat treatment of 1.0Cr-0.5Mo steel". Welding and Materials, vol 4, n 2, April/June 1992, p. 24-27.

TSAI, C. L., AND MASUBUCHI, K. 1979. "Mechanisms of rapid cooling in underwaterwelding". Applied

Ocean Research 1: 99-110.

WAINER, E., BRANDI, S. D., MELO, F. D. "Soldagem-Processos e Metalurgia", led. São Paulo: Edgard Blucher, 1992.494p.

WERNICKE, R., POHL, R. "Underwater Wet Repair-Welding and Strength Testing on Pipe- Patch Joints", 17th International Conference on Offshore Mechanics and Arctic Engineering - OMAE- 1998.

WTIA - "Temper Bead Welding"; Welding Technology Institute Australian, March 2006.

YURIOKA, N. et al. "Carbon Equivalent to Assess Cold Cracking Sensitivity and Hardness of Steel Welds". Nippon Steel Technical ReportNr. 20, Dec. 1982. pp.61-73.

Printed by Books on Demand GmbH, Norderstedt / Germany